Abiogenesis

Part 1

The Essential Chemistry

George E. Parris

2021

Preface

The origin of life has been a question of interest through human history. It has become much more urgent with the search for other living organisms on other planets and in other solar systems (exobiology).

I have already written extensively concerning Evolution, Development and Speciation:

Parris GE. 2009. A Hypothetical Master Development Program for Multi-cellular Organisms: Ontogeny and Phylogeny. 2009. Biosciences Hypotheses 2:3-12. Note: New journal is not widely indexed, via on-line at Science Direct.

Parris GE. 2009. A Hypothetical Master Development Program for Multi-cellular Organisms: Ontogeny and Phylogeny. 2009. Biosciences Hypotheses 2:3-12. Note: New journal is not widely indexed, via on-line at Science Direct.

Parris GE. 2009. Comment on a Hypothetical Master Development Program: Creation and Evolution of Generation Specific Control Keys. 2009. Biosciences Hypotheses Note: New journal is not widely indexed, via on-line at Science Direct.

Parris GE. 2009. A hypothesis: Are pyknons coding units of the Master Development Program. Biosci Hypotheses.;X:355. Note: New journal is not widely indexed, via on-line at Science Direct.

Parris GE. 2010. Developmental diseases and the hypothetical Master Development Program. Med. Hypotheses Mar;74(3):564-73.

Parris GE. 2010. Scope of Medical Implications of the Master Development Program. Med. Hypotheses. 74(5):953.

Parris GE. 2010. Speciation in *Anopheles gambiae* is consistent with the predictions of the Master Development Program. Med. Hypotheses.75:135-136

Parris GE. 2011. The Hopeful Monster Finds a Mate and Founds a New Species. Hypotheses in the Life Sciences. 1(2):1-6.

http://www.hy-ls.org/index.php/hyls/article/download/53/53-164-3-PB.pdf

http://www.amazon.com/Hopeful-Monster-Application-Speciation-Hominidae-ebook/dp/B00MVFG470

Parris GE. 2011.Asymmetric Division and the Immortal Strand Hypothesis. Hypotheses in the Life Sciences. 1(2):52-55. On-line:
http://www.hy-ls.org/index.php/hyls/article/download/55/55-214-3-PB.pdf

http://www.amazon.com/Asymmetric-Division-Immortal-Strand-Hypothesis-ebook/dp/B00MVEJ6A8

Parris GE. 2013. Chimps Descended from Humans
http://www.amazon.com/Chimps-Descended-Humans-George-Parris-ebook/dp/B00HE4RGZC

Parris GE. 2013. Application of a Hypothesis to Speciation in Hominidae. Hypotheses in the Life Sciences. 3(1) on-line:

http://www.hy-ls.org/index.php/hyls/article/download/99/99-346-1-PB.pdf

And as a chemist, I have wanted to tackle the question of a biogenesis (i.e., what happened before living systems began evolving, splintering into different species and developing complex tissues and organs. But there has been little experimental research to draw on for guidance. Moreover, the experimental work that initiated the field appears to me to have misled more than enlightened. Of course, I am thinking about the electric arc experiments by Miller and Urey.

Regardless, this field of study has been dominated by people who seem poorly equipped to propose any realistic chemical mechanisms leading to living systems. My reading of the literature suggest that thought in this area has been led by astrophysicists, marine geologists, statisticians, philosophers and half-baked planetary biologists. The remarkable misadventures at Mono Lake [aptly called "Moon Lake"] and the claim that arsenic can replace phosphorus in DNA (December 2, 2010) was a remarkable excursion into absurdity. It took a couple of years (July 9, 2012) for all the embarrassed people to quietly disengage without admitting boldly that they were complete idiots for believing the claims of Felisa Wolfe-Simon (obo player masquerading a chemist).[1]

By its nature, much of the information I must consider it uncertain, unproven or speculative. Moreover, I simply disagree with some of the popular opinions about what was happening 4 to 5 billion years ago.

George E. Parris, PhD

Gaithersburg, MD 20882

Part 1. I Summarize the **Essential Chemical Reactions**

And

Part 2. will discuss **Formation and Evolution of Rudimentary Cells**

---

[1] Wolfe-Simon F, Switzer Blum J, Kulp TR, Gordon GW, Hoeft SE, Pett-Ridge J, Stolz JF, Webb SM, Weber PK, Davies PC, Anbar AD, Oremland RS. A bacterium that can grow by using arsenic instead of phosphorus. Science. 2011 Jun 3;332(6034):1163-6. doi: 10.1126/science.1197258. Epub 2010 Dec 2. PMID: 21127214.

From her website: "*Dr. Wolfe-Simon did her undergraduate work at Oberlin College and Conservatory of Music where she graduated with a B.A. in Biology (Chemistry) and a B.M. in Music Performance. She went on to earn her Ph.D. in Oceanography from Rutgers University.*"
Source: https://www.felisawolfesimon.com/teaching#!

1.0 Essential Chemistry in an Abiotic World

If we are going to propose abiotic mechanisms leading to living systems, we need to understand what chemicals and energy sources were available and what physical processes controlled their distribution over the period of time during which life arose on earth. I do not think there is any single correct answer regarding how life may have evolved on earth or how it might evolve on other planets. As noted above, as we move from chemistry to biology, there are many random events that could have favored one mechanism over another. *I am not attempting to pick out the actual pathway that was followed on earth, nor and I trying to provide an encyclopedic presentation of all known hypotheses.* What I am presenting is a discussion of some hypotheses, which I think have relatively high probability and illustrate a continuous flow from one system to the next.

1.1 Matter, Energy and the Universe

The Big Bang

Genesis 1:3: *And God said, "Let there be light" and there was light.*

We start with the philosophical. I'm a scientist. Science does a really good job of answering most questions. Specifically: Who, What, When, Where and How? But science ultimately fails at WHY? We may eventually penetrate in time to before the Big Bang…15 billion years ago. That's a "how." But why does anything exist? Why does energy exist? Why do the laws "of nature" exist? In practical terms, I'm a deist…. I believe in God. Indeed, I believe that God created the universe I can observe and all the laws that link it together. Moreover, God does not dabble in human lives by suspending or modifying "natural law." There are no miracles. As a scientist, I venerate God by exploring and trying to understand his creation. When I discover an unknown fact or develop a hypothesis that might be tested to unveil a fact, I am obliged to "witness" to my fellows. The biggest sin I can commit is to misrepresent what I discover. Indeed, every "fact" is a hypothesis.[2]

Bernard Shaw (1856–1950). *Man and Superman*. 1903.
Maxims for Revolutionists:
Never resist temptation: prove all things: hold fast that which is good

---

[2] G. Bernard Shaw: "Every fool believes what his teachers tell him, and calls his credulity science or morality as confidently as his father called it divine revelation."

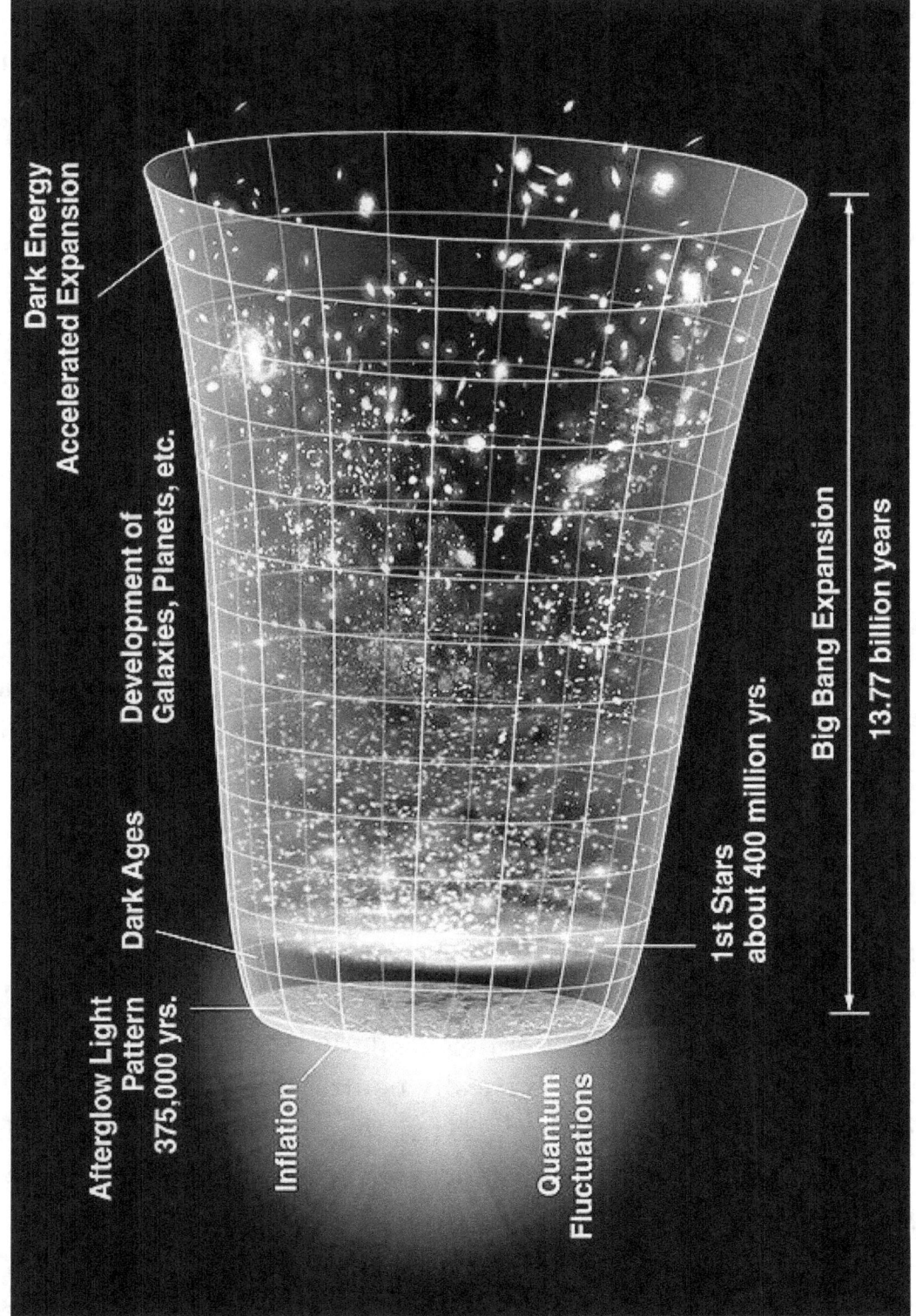

Source: NASA/WMAP Science Team from Wikimedia Commons

There was an enormous amount of energy released into a void of space about 14 billion years ago as electromagnetic radiation. The radiation was characterized by three vectors, which we recognize as charge, magnetic moment and momentum.

As the energy dispersed, it condensed into two fundamental particles we know as protons and electrons. Any other options for capturing the three vectors of energy as "mass, charge and magnetic moment" appear to have been unstable and very transient under the conditions of expansion of the radiation of the energy into space. The two fundamental laws of nature dominating the universe are:

Nothing moves faster than the speed of light in a vacuum

Energy is always conserved instantaneously

With the appearance of mass, the momentum of mass is conserved as mutual attraction (gravity) and the other vector components of mass (charge and magnetic moment) are all conserved instantaneously.

Nucleogenesis how Atomic Nuclei are Formed

In the void of space, the kinetic energy of protons and electrons eventually was dissipated by mutual attractions of opposite charges producing hydrogen atoms ($^1H_1$), which found each other as electrically neutral diatomic molecules. The much weaker gravity slowly acted on the hydrogen molecules causing them to spiral into clouds that brings the molecules into contact and continues to agglomerate more and more hydrogen.

In the center of these clouds the pressure (of the massive depth of mass attracted by gravity) compresses the hydrogen atoms so that their state changes from molecular to metallic: hexagonal close packing of independent spheres (400 GPa;

3,900,000 atm; 58,000,000 psi). Once these are crushed to the limit of the electron wavelength the only option is to fold the electron into the "zero quantum state" (i.e., Bohr atom with n= 0). This state is naturally unstable at ordinary pressure and absent an electric field. We call it a neutron.

The neutron has a mass slightly more than a proton and an electron[3]:

$$Mp = 1.67262192369(51) \times 10^{-27} \text{ kg}$$

$$\mathbf{Me} = 0.00091093837015 \times 10^{-27} \text{ kg}$$

Sum (Electron + Proton) = $1.673532864 \times 10^{-27}$ kg which is less than

$$Mn = 1.67492749804(95) \times 10^{-27} \text{ kg}$$

The difference is $= 0.001395 \times 10^{-27}$ kg per atom

Or 8.3957×10^{-7} kg/mole

Einstein's equation comparison of energy and mass: $\mathbf{E = mc^2}$
Thus, the energy absorbed by creation of the extra mass is

$= (8.3957 \times 10^{-7}$ kg/mole$) \times (3.00 \times 10^8$ m/s$)^2 = 7.556 \times 10^{10}$ J/mole

Or 783 eV/neutron, which represents the energy released during inflation the neutron to the hydrogen atom

$$n \rightarrow [^1H_1 + 783 \text{ eV}] \rightarrow p + beta \quad \text{see footnote}[4]$$

---

[3] The pdv energy term (change in volume) as the hydrogen atom (10^{-10} m) is crushed to a neutron (10^{-15} m) less the electrostatic repulsion must be conserved as mass.

[4] Those of you with a physic background are now pointing at my ignorance and lack of fundamental understanding. Where is the neutrino!! For a full explanation of why I think the neutrino is one of the biggest mistakes in physics consult my book *History of Atomic Theory*

Free neutrons (i.e., neutrons in free space) are thus unstable and readily decay to protons and electrons with a mean lifetime of 879.6 ± 0.8 s. But, the presence of neutrons among protons allows the stabilization of combinations of protons, neutrons and electrons as atoms, which are (at least transiently) stable outside the high-pressure core of massive stars. Thus, when the stars explode, a variety of nuclei are generated (we call them isotopes) and they capture the free electrons.

Briefly, the neutrons in the nuclei of some isotopes are stabilized, by the fact that the 783 eV produced by their expansion is insufficient to expel an electron from the atom against the electrostatic attraction of the nucleus. Thus, although some neutrons are likely continually expanding to hydrogen in the nuclei of "stable" isotopes, the electrons do not have sufficient energy to escape from the nucleus and thus return, to the "neutral doublet."[5]

Isotopes and Elements

The isotopes produced in the explosion of the first generations of giant stars covered a wide range of possibilities, but most of them were only fleetingly stable without the crushing pressure of the star.

It turns out that the nuclei become more stable out to an isotope we call iron-56. It contains 26 protons and 20 neutrons and thus formed a stable element with 26 electrons.

––––––––––––––––––––––

(April 18, 2019) sold on Amazon where I explain my skepticism for your benefit. Here I'm not going to distract the biologists with your confusion.   The biologists need to know that equipartition of momentum between the proton and electron means and conservation of energy means that most of the energy (783 eV less the ionization potential of hydrogen) will end up with the much less massive electron propelling it to high velocity as a beta ray.

[5] I actually envision the nucleus much like J.J. Thomson's model of the "plumb pudding" atom or Bohr's atom with the principal quantum number $n = 0$.

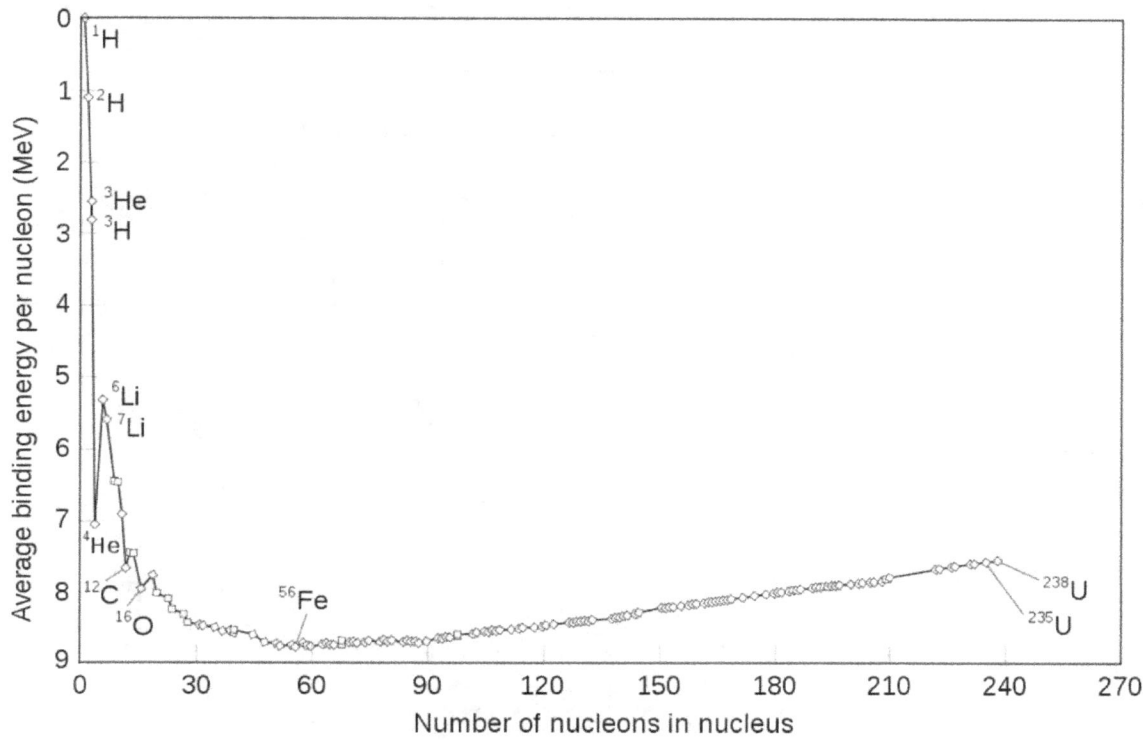

Source: http://large.stanford.edu/courses/2010/ph240/hamerly2/images/f1big.png

Nuclear combinations with equal or approximately equal numbers of protons and neutrons tended to be stable (or have indefinitely lifetimes on our timescale) out to this number of protons. If there are too many neutrons, the isotope is unstable relative to neutron inflation (as discussed above). We call the escaping electrons *beta particles (or beta rays* because of their high kinetic energy). On the other hand, if there are too many protons, the nuclei tend to split out the stable *alpha particle* (helium nucleus 2 protons with two neutrons). After about 12 billion years, we have the mix of isotopes depicted below. It is apparent that nuclei with an excess of protons tend to split off helium (alpha particles) more readily than nuclei with an excess of neutrons tend to lose beta particles. Of course, loss of a beta particle, leaves behind a nucleus with a potentially destabilized number of protons and loss of a beta particle is frequently followed by loss of an alpha particle.

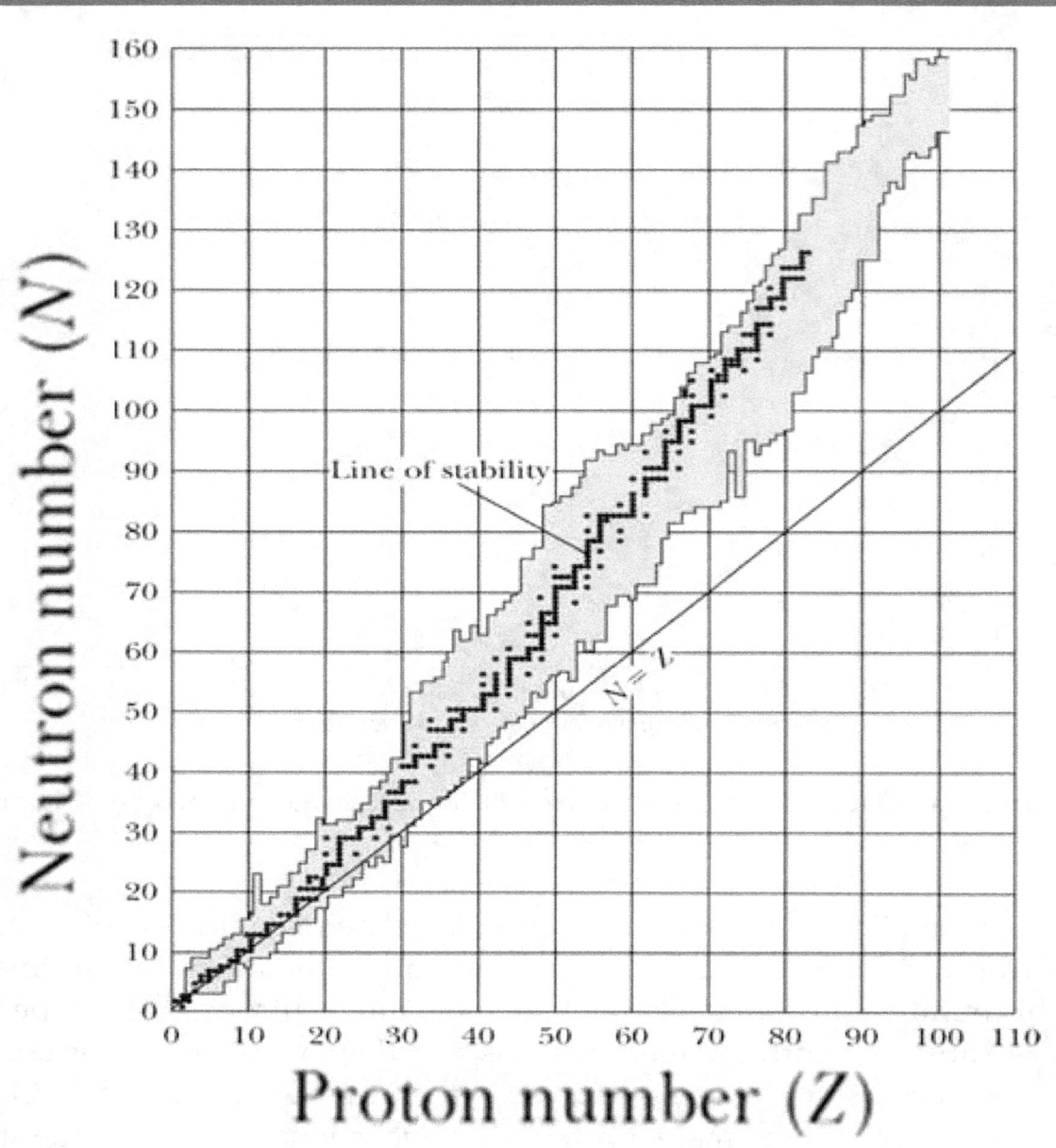

Source:
https://classconnection.s3.amazonaws.com/799/flashcards/1528799/png/nuclear_stability1355
021984570.png

Fission of Heavy Isotopes and Nuclear Radiation

This is all interesting to physicists but is only secondarily of interest to biologists.

It is relevant that no isotopes with more than 92 protons are stable on the time scale of billions of years. And none of the isotopes with 43 protons are stable enough to survive on that timescale. They are observed in the atomic spectra of giant stars, but outside of that environment their longest half-lives are only in the millions (10^6) of years. It turns out that isotopes we call Uramium-238, Uranium -235, Thorium-232 and Potassium-40 have half-lives on the order of about a billion (10^9) years and still persist today in significant quantities. Indeed, the energy released by the decay of these isotopes (and the unstable isotopes that result from their decay) is what keeps the core of the earth hot and molten. Without this continuing release of energy, the earth and similar planets would have cooled to solid rocks long ago (as calculated by Lord Kelvin in the 1800s). Keep in mind that with a half-life of a billion years, the level of radiation release and heating was twice what it is now a billion years ago; such that if we go back 4 billion-years, we go to 2, 4, 8, 16-times the current heating rate and the earth was a lot more fluid. This has geological consequences that we will discuss later.

The other important point is that isotopes surround themselves with negatively charged electrons. These are called elements. Th relevant point is that the electron configuration (determined by the number of protons…not the total number of nuclear particles protons + neutrons) determine the chemistry of the isotopes and thus, we lump all the *atoms that have the same chemistry [same number of protons and electrons] together as an element*…regardless of their mass or stability. Thus, we talk about the isotopes of an element (not the elements of an isotope). The chemists, have thus, organized science around the elements, which they have characterized in the periodic chart in which the *mass of the elements is given as an average of their naturally occurring isotopes*.

Periodic Table of the Elements

Source: https://www.activityshelter.com/wp-content/uploads/2017/10/print-periodic-table-of-elements-free-picture.png

Allotropes, Compounds and Polymers

The primordial gas, stable isotopes and electrons from exploding giant stars expand into space converting kinetic energy into potential energy of gravitational, electrostatic and magnetic attractions. The residual kinetic energy can be equated to "heat" and the universe cools to the point that the relative kinetic energy of the particles is on the scale of the attractions of nuclei and electrons. Atoms of various elements are condensed and further cooling allows chemical bonding to occur.

The condensation of electrons and isotopes does not happen randomly. Quantum mechanical restrictions require that the electrons, which have very little mass, display dual particle-wave properties. When constrained to a volume of space defined by proximity to a positively charged nucleus, the wave form of the free electron must be accommodated. The result is that the independent electrons[6] must occupy only *specific volumes of space around the nucleus* (determined by their mutual electrostatic attractions); we call these *orbitals*.

But the individual electrons each has negative charge and a magnetic moment. As a result, the electrons to the extent possible pair up (i.e., form *electron pairs*) in which their mutual electrostatic repulsion is balanced by magnetic attraction ($\uparrow\downarrow$) which neutralizes their magnetic moments. Inevitably, some atoms have unpaired electrons in the least stable (outer) orbitals. In some cases, this is because the nuclei have odd numbers of protons and in others the volume of the orbital is large enough that the energy reduction from forming magnetic-cancelling pairs ($\uparrow\downarrow$) is not strong enough to ensure that it overcomes the mutual electrostatic repulsion of the electrons.

In these cases, the potential for electron-pairing between/among separate atoms is answered at relatively low cosmic temperature (e.g., below a few thousand degrees Celsius). From this interaction, simple agglomerations of atoms can occur through "*chemical bonding*." Interestingly, chemical bonding can take on a

---

[6] "Independent electron" means not associated with the nucleus in the form of neutrons.

variety of forms. Hydrogen (under low pressure) provides an example for the simple two-electron single bond:

> Two hydrogen atoms approach one another to a point where their electrons can form a pair. The electron density lies primarily between the two nuclei and produces an electrostatically more stable *diatomic molecule.*

$$H + H \rightarrow H\text{-}H + energy$$

We call this an *allotropic form* of hydrogen. Overall, seven elements prefer this allotrope (diatomic molecule: hydrogen, oxygen, nitrogen, fluorine, chlorine, bromine and iodine).

A variety of ways of accomplishing this sort of paring are available when the array of different atoms is presented (see below). But most elements actually follow a different path: They share electrons among a large number of atoms. The atomic nuclei with a core of closely bound electrons form a *cation* (positively charged ion of one or more atoms) that float in a sea of paired electrons. The electrostatic repulsion among the pair of electrons actually expands the array of cations (which represent most of the mass of the material) to the point that they can slip past one another…deforming the array…with relatively little effort. We call these allotropes *metals*. Metals are (i) *malleable*, (ii) melt into mobile fluids and (iii) conduct electricity and heat very well because the electrons are free to move among the mass of cations.

The most interesting type of allotrope is the *polymer* or *polyatomic molecule*. For example, sulfur atoms (under what we consider ambient conditions of temperature and pressure) form eight-member rings of sulfur atoms.

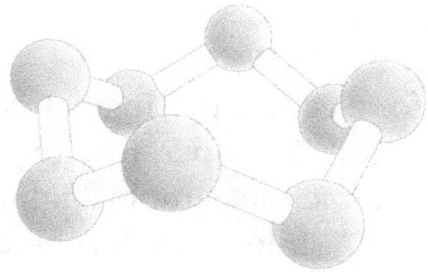

Source: https://elasticbeanstalk-us-east-1-939261147616.s3.amazonaws.com/s8-01.jpg

Carbon occurs in many allotropes: most commonly diamond, graphite and soot all of which are *polymeric* (C_n). But in recent years, interest has turned to specific molecular allotropes of carbon (e.g., C_{60}):

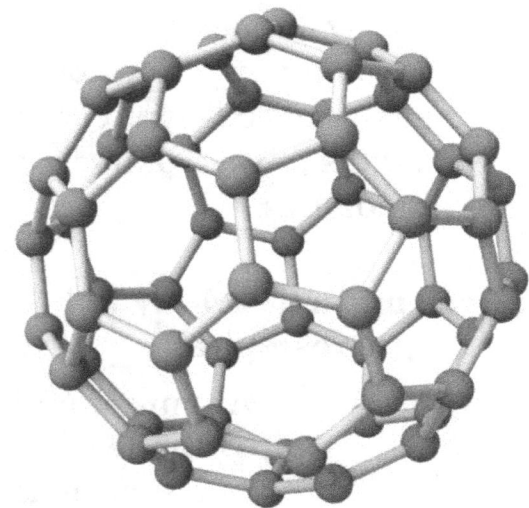

Source: https://www.c60labs.com/wp-content/uploads/2018/03/60-Molecule.jpg

The final type of allotrope (compounds formed of only one element) are the simplest. The inert (mon-atomic) gases (helium, argon, neon, krypton, xenon and radon). These elements have no unshared electrons and have no tendence to form (electron sharing) *covalent bonds*. Some of them however can be oxidized (have electrons removed) to make ions with odd numbers of electrons that will form compounds with other elements.

Most compounds of interest involve combinations of different element. For example, many elements from compounds with hydrogen: LiH, MgH_2, BH_3, CH_4, NH_3, H_2O and HF). From here we could launch into a long discussion of chemistry concerning the polarity and strength of each type of bond and the reasons for various elements having various bond angles and reactivities. But I am writing for an audience that should be well versed in general and organic chemistry. *Organic chemistry* being that interesting area involving carbon compounds primarily with hydrogen, oxygen and nitrogen, which are the target of this book.[7]

The Volatile Compounds

The spiral flow of elements from the first-generation stars that were drawn together by gravity was dominated by hydrogen molecules (H_2) and other molecules and atoms based primarily on their atomic abundance (not their chemical stability under ambient conditions). The abundance of molecules depended primarily on the frequency of collisions of the atoms in space. Surveys have been conducted and the physics of nuclear fusion has been consulted to identify relative abundances.

		Ten most common elements in the Milky Way Galaxy estimated spectroscopically		
Z	Element	Mass fraction (ppm)	Atomic abundance (divide mass by atomic mass)	Relative Atomic abundance
1	Hydrogen	739,000	739,000	1.000
2	Helium	240,000	60,000	0.081
8	Oxygen	10,400	650	0.00088
6	Carbon	4,600	383	0.00052
10	Neon	1,340	67	0.000091
26	Iron	1,090	19	0.000026
7	Nitrogen	960	69	0.000093
14	Silicon	650	23	0.000031
12	Magnesium	580	24	0.000032
16	Sulfur	440	14	0.000019
	Total	999,500		

Source: https://i.stack.imgur.com/sT7Ue.png

---

[7] If you are not familiar with chemistry, you will have to take my word for it in the following discussions.

Thus, if we start looking for diatomic combinations *in space,* we would expect lots of HX molecules: H_2, HO, HC, HN, HS. This is confirmed in the current atmosphere of Jupiter where H_nX compounds (X = C; O, S, Se; N, P, As; etc.) predominate. We will disregard the inert gases (He and Ne) and the metal hydrides that are not stable under ambient conditions. The important molecules for our story are the ones that are reactive and subsequently form tri- and tetra-atomic compound through subsequent collisions with hydrogen atoms, hydrogen molecules or less common mono and diatomics (especially the relatively common O and O_2). It is relevant that because of the likely high kinetic energy of the colliding monatomic and diatomic species, and the requirement of conservation of momentum, only molecules that can dissipate much of that kinetic energy in molecular rotations and vibrations are likely to form more complex molecules and molecular fragments. Consulting a table of bond energies (i.e., energy needed to break diatomic bonds) the following combinations look most likely *in interstellar space* which is limited by collision probability.

High Stability Bonds		Relative Abundance of combinations $pA \times pB =$	Probability Index Bond Energy X Abundance of atoms	
Combination	Bond Energy			
CC	965	2.7×10^{-7}	2.6×10^{-4}	10^{-6}
NN	945	8.6×10^{-9}	8.1×10^{-6}	10^{-8}
CN	891	4.8×10^{-8}	4.3×10^{-5}	10^{-7}
CO	732	4.6×10^{-7}	3.3×10^{-4}	10^{-6}
NO	631	8.1×10^{-8}	5.1×10^{-5}	10^{-7}
OO	498	7.7×10^{-7}	3.9×10^{-4}	10^{-6}
HO	460	8.8×10^{-4}	4.0×10^{-1}	10^{-3}
HH	436	1	436	1
CH	410	5.2×10^{-4}	2.1×10^{-1}	10^{-3}
NH	390	9.3×10^{-5}	3.6×10^{-2}	10^{-5}
SH	340	1.9×10^{-5}	6.5×10^{-3}	10^{-5}
CO_2	1600	4.0×10^{-10}	6.5×10^{-7}	10^{-9}
NCO	1360	4.3×10^{-11}	5.8×10^{-8}	10^{-10}
HCN	1301	4.8×10^{-8}	6.2×10^{-5}	10^{-7}
HCO	1158	4.6×10^{-7}	5.3×10^{-4}	10^{-6}
H_2O	920	8.8×10^{-4}	8.1×10^{-1}	10^{-3}
CH_2	820	5.2×10^{-4}	4.3×10^{-1}	10^{-3}
SiO_2	800	2.4×10^{-11}	1.9×10^{-8}	10^{-10}
H_2S	680	1.9×10^{-5}	1.3×10^{-2}	10^{-4}
OCCO	2078	2.1×10^{-13}	4.4×10^{-10}	10^{-12}
CH_4	1640	5.2×10^{-4}	8.5×10^{-1}	10^{-3}
H_2CO	1552	4.6×10^{-7}	7.1×10^{-4}	10^{-6}
NH_3	1170	9.3×10^{-5}	1.1×10^{-1}	10^{-3}
HNCO	1820	4.3×10^{-11}	8.7×10^{-8}	10^{-10}
C_2H_2	1785	2.7×10^{-7}	4.8×10^{-3}	10^{-5}
C_2H_4	2605	2.7×10^{-7}	7.0×10^{-4}	10^{-6}
C_nH_{2n+2}	$(2n +2) \times 410$	$(5.4 \times 10^{-4})^n$	---	$< 10^{-9}$

Once interstellar gases have condensed into planetary atmospheres, we can assume that all the molecular fragments combine driven primarily by bond energy to saturate the valances of all elements consuming reactive H_2 and O_2.

High Stability Bonds		Initial Relative Abundance	Likely Products	
Reactants	Reactivity			
CC	Very high	10^{-6}	C_2H_4/C_2H_2O	10^{-6}
CN	High	10^{-7}	HCN	10^{-7}
CO	Moderate	10^{-6}	CO/CO_2	10^{-6}
NO	High	10^{-7}	NO_2	10^{-7}
OO	High	10^{-6}	H_2O	10^{-6}
HO	Very High	10^{-3}	H_2O	10^{-3}
HH	Moderate	1	H_2	1
CH	Very High	10^{-3}	CH_4/CH_3OH	10^{-4}
NH	Very High	10^{-5}	NH_3	10^{-5}
SH	Moderate	10^{-5}	H_2S/SO_2	10^{-5}
NCO	Moderate	10^{-10}	HNCO	10^{-10}
HCO	High	10^{-6}	$H_3CCOH/H_2CO/HCO_2H$	10^{-6}
CH_2	Very High	10^{-3}	H_2CO	10^{-7}

Which results in an initial (pre-solar) planetary atmosphere predominantly of

$H_2 >>> H_2O, CH_4, >> H_3COH, NH_3 > H_2S, SO_2, H_2CO, > CO, CO_2, HCN, NO_2$

The light elements (primarily H_2) diffused rapidly to the gravitational center of the spiral and built up the sun. The heavier elements (primarily iron and other metals with stable oxides of Mg and Si) formed the cores of planets.
As discussed in more detail below, as soon as the solar hydrogen fusion cycle was initiate, the lighter (non-condensable components) primarily H_2, CH_4, and other hydrocarbons, CO, CO_2, and HCN were swept away from the inner planets and accumulated in the outer planets.

The Phosphorus Delma

Phosphorus is not a particularly common element. Phosphorus-31 is the only stable isotope. Indeed, most of the phosphorus isotopes undergo beta decay to make sulfur or beta decay followed by loss of a proton or alpha etc. to make magnesium, silicon or aluminum. There is about 1 phosphorus atom for every sulfur atom in the universe.[8] Thus, phosphorus that one might expect to be rather common is actually fairly rare in the universe and may be heterogeneously allocated to specific solar system and planets. There is evidence that the ratio of phosphorus to iron varies substantially among supernova.

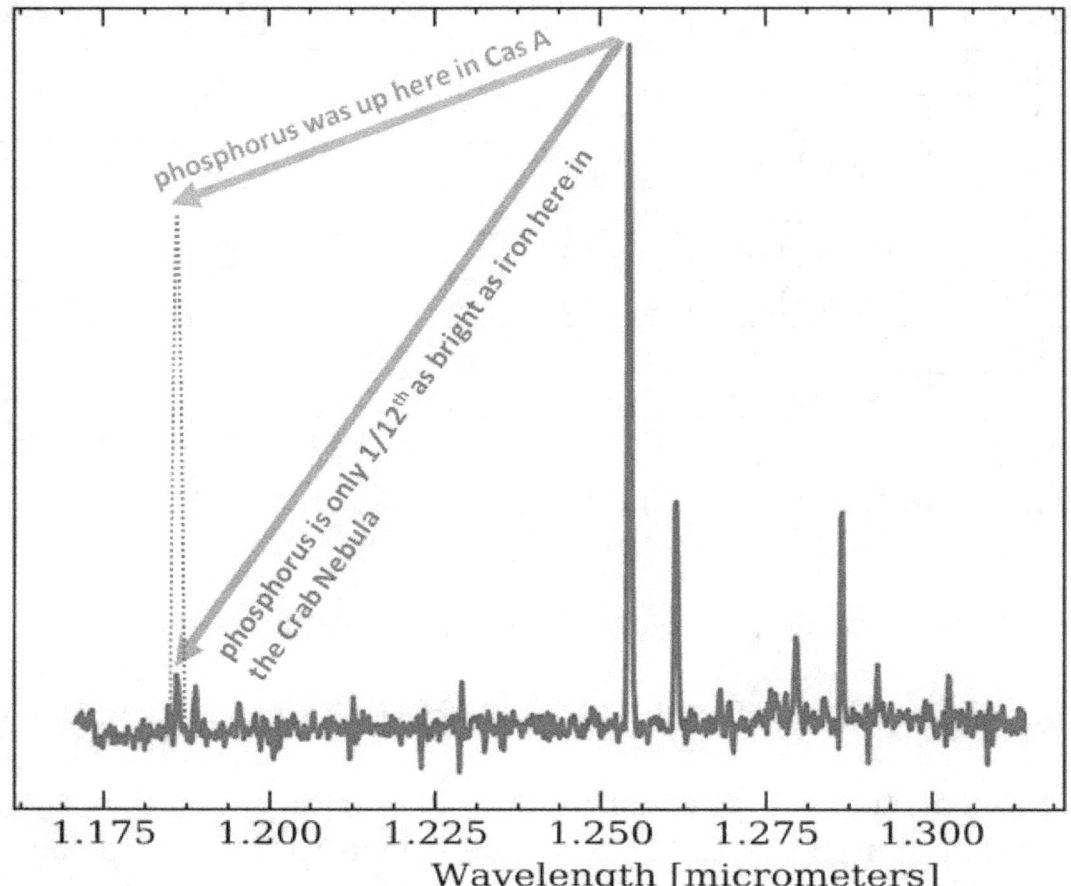

Source: https://newatlas.com/phosphorus-shortage-life-alien/54080/#gallery:2

But phosphorus is fairly common on earth:

---

[88] https://periodictable.com/Properties/A/UniverseAbundance.v.log.html

Elemental Composition of Earth					
Crust				overall	
Element	Percentage Weight (Atomic)	Atomic Oxygen Consumption[9]		Element	Weight Percentage
Oxygen	46.1 (2.88)			Iron	34.6
Silicon	28.2 (1.01)	2.00		Oxygen	29.5
Aluminum	8.23 (0.305)	0.20		Silicon	15.2
Iron	5.53 (0.099)	0.15		Magnesium	12.7
Calcium	4.15 (0.104)	0.10		Nickel	2.4
Sodium	2.36 (0.103)	0.05		Sulfur	1.9
Magnesium	2.33 (0.097)	0.10		All others	3.7
Potassium	2.09 (0.054)	0.02			
Titanium	0.565 (0.012)	0.02			
Hydrogen	0.14 (0.14)	0.07			
Phosphorus	0.105 (0.003)	0.001			
All others	0.174				
Handbook of Chemistry and Physics, 89th ed. (Boca Raton, FL: CRC Press, 2008–9), 14–17.					

This has implications regarding the abiogenesis on earth relative to other similar planets. If life evolved everywhere using the same chemistry, it may have been much slower or more limited on other planets. Or, life may have evolved with different chemistry. Note that phosphorus and potassium are essential for DNA and RNA.

1.2 The Sun and Solar System

Collapse of hydrogen (^{1}H) atoms to neutrons[10] that are stabilized as deuterium (^{2}H) followed by fusion of deuterium (^{2}H) to make helium (^{3}He), which continues to stable helium (^{4}He) releases an enormous amount of energy deep in the sun. This energy is transmitted as thermal energy to the surface, where the random

---

[9] Based on the empirical formula of the oxides, what relative amount of oxygen is associated with each element. For example: Al_2O_3, CaO, K_2O.

[10] I am not convinced that "neutrinos" exist. I believe they will go the way of "caloric."

collisions of atoms cause the release of a continuous spectrum of photons. This release of photons is a manifestation of the simultaneous conservation of momentum and energy.

Sunlight and Electromagnetic Radiation

Based on the pattern of solar radiation (i.e., the wavelength of maximum luminosity), we know (from Wein's law) that the average surface temperature of the sun is nearly 6,000 °K.

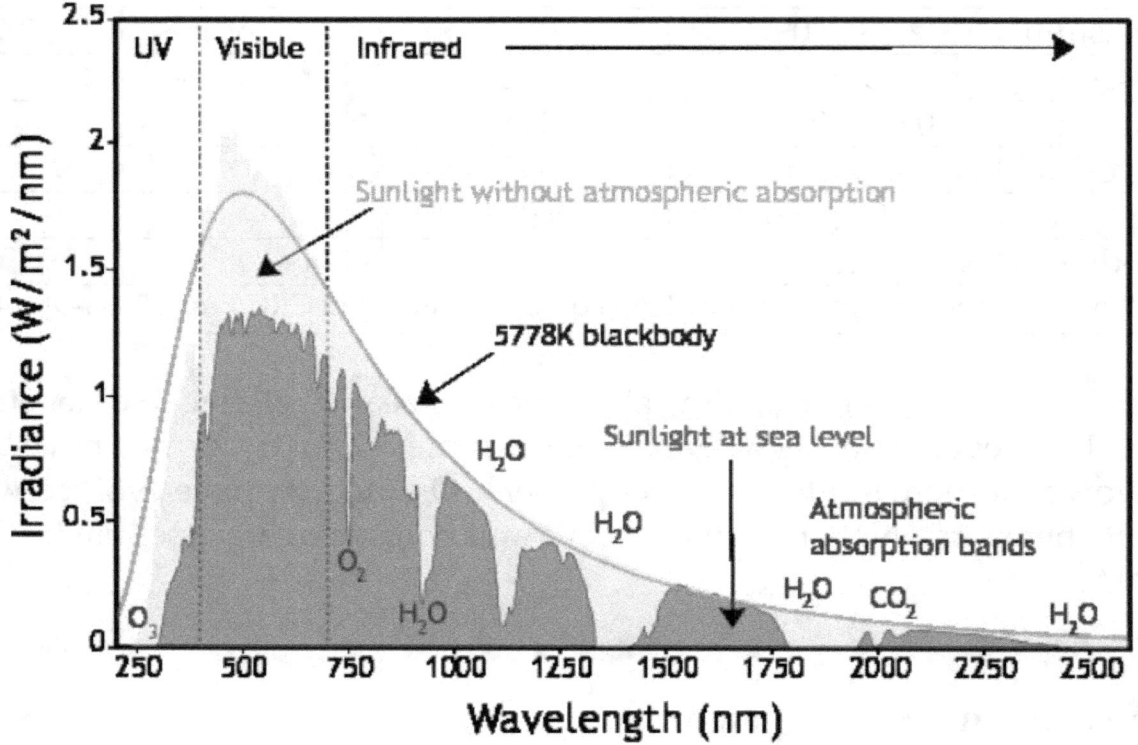

Source: https://www.researchgate.net/figure/The-spectrum-of-solar-radiation-in-wavelengths-near-the-visible-spectrum-Notice-the_fig2_327521724

In addition, the sun releases a spray of electrically charged particles, which first gain our attention as the northern lights (*Aurora Borealis*):

Source: https://i.pinimg.com/736x/16/79/11/167911f93a582356e7786146767c6970.jpg

And recently observed as a "space hurricane":

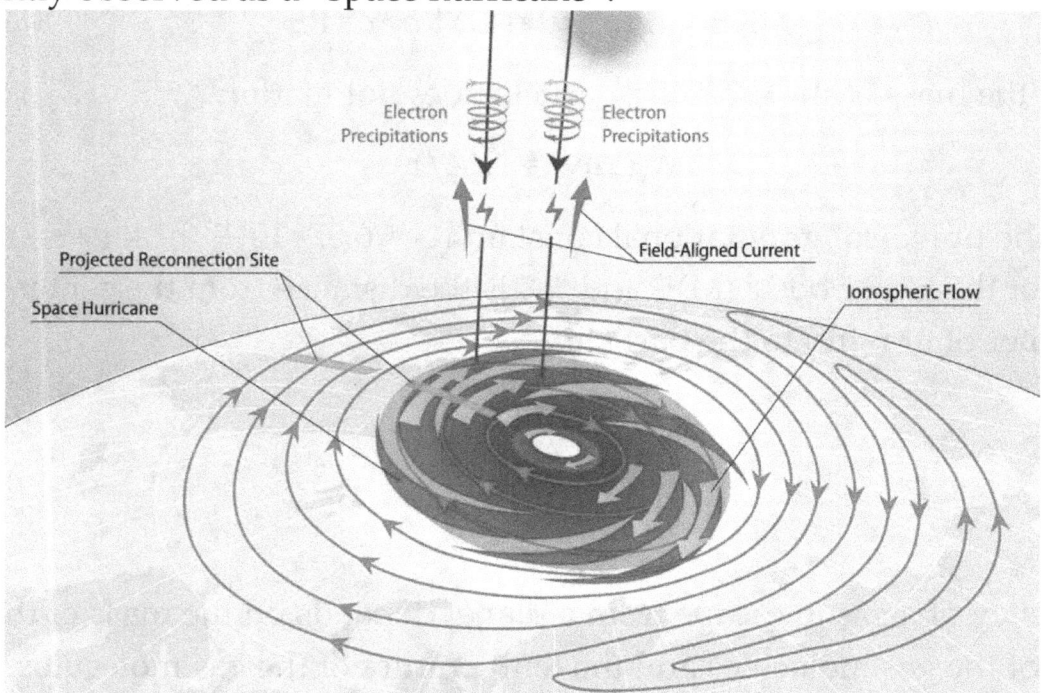

Source: https://www.wpri.com/wp-content/uploads/sites/23/2021/03/space-hurricane.jpg

The charged particles are electrons, protons, and alpha particles (0.5 and 10 keV) with occasional heavier nuclei. They are guided to the poles by our magnetic fields, and otherwise would be dangerous to life on earth.

Escape Velocity[11]

Once the planetary cores were formed from metals and oxides (primarily iron and silicates), they began to accumulate lighter elements be gravitational attraction. The disposition of these atoms and molecules was determined by the escape velocity associated with each planet and the temperature (which determines the velocities) of the gas molecules.

Equating the kinetic energy (KE) of a body (it does not matter what the mass of the body is… it could be a boulder or hydrogen molecule) to the potential energy (PE) needed to remove it to infinite distance from the earth (in general a larger body of mass "M") allows the derivation of the equation of the escape velocity:

$$KE = (½) \, mv^2 = GMm/r = PE$$

Note that the mass of the escaping particle does not matter:

$$(½)v^2 = GM/r$$

"GM" is the universal gravitational constant ($G \approx 6.67 \times 10^{-11}$ m^3·kg^{-1}·s^{-2}) times the mass of the larger body (M)[12] and "**r**" is the distance from the center of the earth (center of gravity) to the particle.

$$v_e = \sqrt{\frac{2GM}{r}}$$

The tendency of a gas to escape from a planet depends on the mass of the planet, the mass of the gas molecules and the temperature of the gas molecules. In the absence of any alternative argument, I take the position that the initial masses of

---

[11] I am providing this section to clarify why earth's primordial atmosphere contained very little molecular hydrogen (H_2) or methane (CH_4) and how high temperatures also limited the final mass of the atmosphere by loss of inert gases and nitrogen (and thus determined the surface atmospheric pressure).

[12] Currently for earth GM = 3.99 x 10^5 km^3/s^2.

the planets (composed of iron, trace metals and silicates) probably did not vary by more than two orders of magnitude (10^{23} to 10^{25} kg) at the time the solar fusion reaction began. And each planet probably gathered around it a diffuse atmosphere of molecules still spiraling towards its core. I similarly assume that the cores themselves were of similar temperature (e.g., 6000 $^\circ$K) of molten iron from the kinetic energy of collisions and radioactive decay.

The question then is what effect the solar radiation heating may have had on the diffuse atmospheres.

Planet	Mature Mass[13] kg x 10^{23}	Distance from the Sun (AU)	Solar Radiation Relative[14]	Radiant Temperature[15] ($^\circ$K)
Mercury	3.3	0.39	6.6--	1980
Venus	48.7	0.72	1.9--	570
Earth	59.7	1.00	1.00-	300
Mars	6.4	1.52	0.43-	129
Jupiter	18986.0	5.20	0.04-	12[16]
Saturn	5684.6	9.54	0.01-	4
Uranus	868.1	19.18	0.003	4
Neptune	1024.3	30.06	0.001	4

If we plug M (10^{24} kg) and r (10^3 km) into the escape velocity equation

$$E_{scape} = [2\,(6.67 \times 10^{-11} \text{ m}^3 \cdot \text{kg}^{-1}\, \text{s}^{-2} \times 10^{24} \text{ kg})/10^6 \text{ m}]^{1/2}$$

---

[13] I believe that all the planets had metallic cores of similar mass when the sun ignited in nuclear fusion.

[14] Inverse square law for radiant energy.

[15] These temperatures essentially assume no atmosphere and are driven exclusively by sunlight scaled to earth at 300°K. The average surface temperature of earth is reported as about 290°K, but the temperature in the deserts can reach 330°K.

[16] The lower atmospheres of the gas giants are currently warmed by ongoing atmospheric compression to about 200°K. The energy density of space is about 4°K

$$= [2\,(6.67 \times 10^{-11}\ m^2 \times 10^{24})/10^6\ s^2]^{1/2} = 11\ km/s$$

Keep that figure (11,000 m/s) in mind and calculate the *root mean square* velocity of a gaseous hydrogen molecule at the radiant temperature using this equation:

$$v_{RMS} = \sqrt{\frac{3RT}{M}}$$

Where R = the gas constant (8.314 J/(mol °K) or 8.314 kg m^2/s^2/(mol °K)), T = temperature (°K) and M = molecular weight of hydrogen molecules (2 x 10^{-3} kg/mole).

Planet	Radiant Temperature[17] (°K)	V_{RMS} (m/s)			
		H$_2$ (2)	Methane (CH$_4$, 16)	Nitrogen (N$_2$, 28)	Phosphine (PH$_3$ and SH$_2$, 34)
Mercury	1980	5000	1800	1300	1240
Venus	570	2700	940	710	650
Earth	300	1900	680	520	470
Mars	129	1300	450	340	310
Jupiter	12[18]	390	140	100	90
Saturn	4	220	80	60	50
Uranus	4	220	80	60	50
Neptune	4	220	80	60	50

With the assumption that all the planet cores were similar to earth initially, with an escape velocity of about 11,000 m/s, it is clear that initially lighter gases would be stripped from the inner planets and be caught by Jupiter which would grow rapidly in mass by accumulation of gases into a dense atmosphere. The inner planets would never have the opportunity to form a dense atmosphere of

---

[17] These temperatures essentially assume no atmosphere and are driven exclusively by sunlight scaled to earth at 300°K. The average surface temperature of earth is reported as about 290°K, but the temperature in the deserts can reach 330°K.

[18] The lower atmospheres of the gas giants are currently warmed by ongoing atmospheric compression to about 200°K. The energy density of space is about 4°K

these lighter gases although SO_2 (mass 64) and H_2SO_3 (sulfurous acid, mass 82) may be volatile and relatively slow moving on Venus. The table below gives an idea of the fractions of various gases that would exceed the escape velocity at about 300°K.

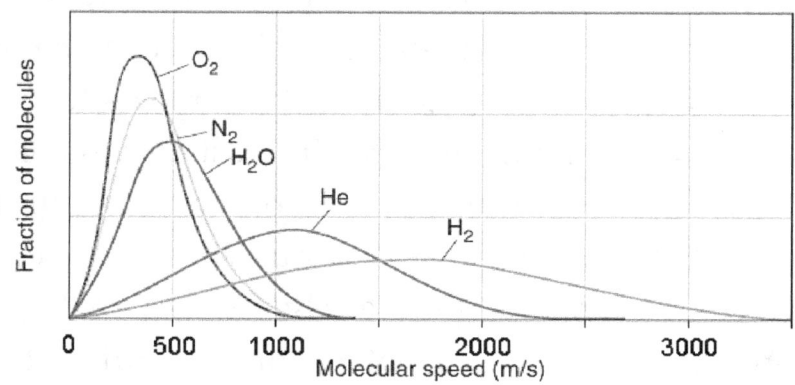

Source: http://www.chem.ufl.edu/~itl/2045/lectures/lec_d.html

Escape Velocity from Surface (in km/s)	
Mercury	4.25
Venus	10.36
Earth	11.19
Mars	5.03
Jupiter	60.20[19]
Saturn	36.09
Uranus	21.38
Neptune	23.56

Today, the gas giants have gained so much mass that even with compression heating (to 200°K), they retain even light molecules.

As an empirical data point, it is clear that nitrogen (N_2, a stable non-polar, inert gas with mass 28) has been retained in our atmosphere in quantity. And I do not see any mechanism for release of nitrogen from the core. Thus, I assume that the rate of loss of nitrogen from earth has been largely balanced by the reception of nitrogen (N_2) from Mercury and Venus.

Similarly, earth may have received phosphine (PH_3) and hydrogen sulfide (H_2S) (molecular mass 34) from the inner planets.

---

[19] Based on current mass and radius. Initially, the masses of the outer planets were probably similar to earth.

Water and carbon dioxide may also have been received from inner planets. But retention of CO_2 (molecular mass 44) and water[20] (with NH_3, which is highly soluble in water) on earth as liquids, solids and mineral hydrates (as opposed to gases) probably account for the abundance of water on earth.

Thus, the solar system model that I envision involves an outward flow of gases from the inner planets to the outer planets. Earth happens to be positioned to gain as much nitrogen (N_2) from the inner planets as it loses to the outer planets while retaining water and ammonia. And the interesting possibility that the earth could be enriched in phosphorus by transport of phosphine (PH_3) from Mercury and Venus might resolve the phosphorus dilemma noted above. It is relevant that on Mercury several metals (Hg, Mg, Cd, Zn, Sn, Pb, Sb, Bi, Al) have significant vapor pressure and may be ablated from the surface to be deposited on cooler planets.

1.3 The Earth's Physical Evolution

In this section, we will set the stage for pre-biotic chemistry on earth. These reactions occurred in the condensed atmosphere, on the surface and in the aquatic environments of the earth. They depended upon the energy balance of the earth and the form that energy took. In addition, physical processes that made specific reactions plausible and brought pre-biological compounds together are discussed.

I have summarized some of the important milestone in the pre-biological evolution of the earth in the table below. In this book, I will focus on events between the stabilization of the earth's crust (3.9 billion year ago) and the rise of

---

[20] Water and ammonia are not as massive as CO_2, but they are polar and form strong hydrogen bonds ($H_2O...H-OH$) which greatly reduces their volatility.

complex, multi-clonal organisms (0.525 billion years ago). But it is desirable to put this period in some context.

Key Milestone in the Earth's History	
billion years ago	Events
10 to 14	The Big Bang and formation of the Universe
4.57	Our sun and solar system established
4.55	Formation of the earth and moon
3.9	Earth's crust stabilized
3.6 to 3.7	Life appeared on earth
3.5	Photosynthesis (green plants)
0.525	Oxygen atmosphere (aerobic respiration)
0.525	Vertebrates appeared on earth
0.165	Mammals appeared on earth
0.050	Primates appeared on earth
0.006	Most Recent Surviving Common Ancestor of Humans
0.004	Modern humans appeared on earth
0.000140	Biological Eve (common female ancestor)
0.000060	Biological Adam (common male ancestor)
0.000012	Agriculture

The Core, Crust and Continents

The early history of the universe was dominated by the formation elements through nuclear fusion and neutron capture processes. It turns out that iron and nickel have the most stable nuclei. Iron and nickel that form metallic allotropes with low vapor pressure condensed from the gas phase into liquids and formed high density planetary cores. These cores were generally coated with silicon oxide (silicate) polymers, which have relatively low density, low volatility and are stable at high temperature. In the crust of the earth *today*, oxygen and silicon are the most abundant elements. Silicates make up the silicate backbone of igneous rocks; with aluminum, iron, calcium, sodium, potassium, magnesium and titanium as cations needed to balance the charges in the silicate polymers. Hydrogen (in the form of water and carbon compounds) and phosphorus round out the top dozen elements in the earth's crust.[21] Phosphorus may have been enriched in the earth's crust by condensation of PH_3 drifting outward from Mercury and Venus. In contact with silicates, phosphine would be incorporated as phosphosilicates and pyrophosphates.

As the earth cooled, the silicates and phosphates acted like flux or slag on molten metal drawing trace elements from the ion/nickel planetary cores and collecting sodium, potassium, magnesium, and calcium oxides from the disk of atomic gases circling the sun. Anionic fluoride, chloride, bromide, oxides, sulfide and traces of other elements also were accumulated in the silicate/phosphate dross along with reduced forms of non-metals (H_2O and H_2S). Depending upon the temperature, pressure and cations present, silicates form a wide range of mineral types. Water released from the silicates at this time was a gas as the earth's surface was above 100°C.

---

[21] Carbon is an abundant element in the earth's crust today, but initially it must have been as CO_2 in the atmosphere because carbonates are unstable at high temperature. Carbonate rock (limestone) is actually post-biotic.

Very Important Pyrophosphate Rock

As the silicate mantel cooled, phosphate separated as pyrophosphates (e.g., phosphorus oxides (P_4O_{10}))[22] trapped within silicates (Mysen and Cody, 2001; London et al. 2001).[23] In addition to the simple cluster compound P_4O_{10} (density 2.39 g/cc) polymeric sheets with higher density can be formed (especially under pressure). The "O-form" has density of 3.05 g/cc and melts at 580°C, and the "O'-form" has a density of 2.72 g/cc:

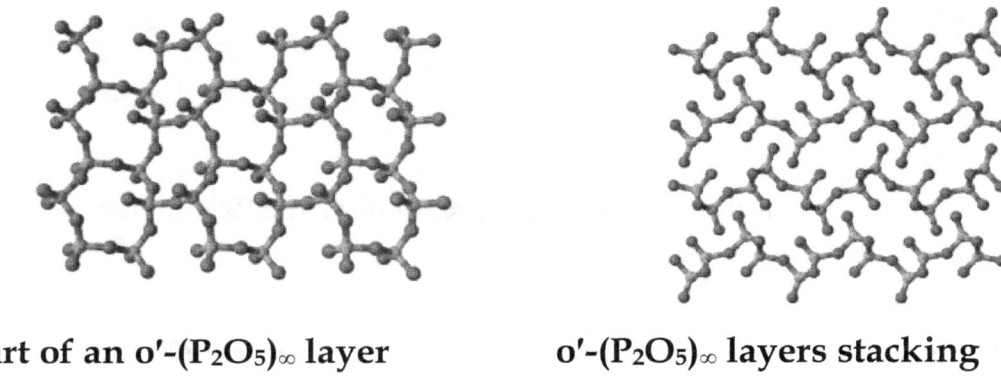

part of an o'-(P_2O_5)$_\infty$ layer   **o'-(P_2O_5)$_\infty$ layers stacking**

Source: http://en.wikipedia.org/wiki/Phosphorus_pentoxide

These polymers are similar to silicates and likely formed in mixed phases with silicates and phosphosilicates.

Weathering of these crustal minerals released active high-energy phosphates over a long period of time. Some of the resulting phosphates readily precipitates

---

[22] Y. Yamagata, H. Kojima, K. Ejiri and K. 1982. Inomata. AMP synthesis in aqueous solution of adenosine and phosphorus pentoxide. *Origins of Life* 12. 333-337.

[23] Mysen BO and Cody GD. 2001. Silicate-phosphate interactions in silicate glasses and melts: II. quantitative, high-temperature structure of P-bearing alkali aluminosilicate melts. *Geochimica et Cosmochimica Acta.* 65(14):2413-2431.

London D, Wolf MB, Morgan GB, Garrido MG. 1999. Experimental Silicate–Phosphate Equilibria in Peraluminous Granitic Magmas, with a Case Study of the Alburquerque Batholith at Tres Arroyos, Badajoz, Spain. *J Petrology.* 40(1):215-40.

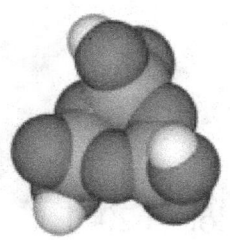 as the poorly soluble mineral apatite and the remainder is in the stable form of polyphosphates. Trimetaphosphoric acid and other pyrophosphates and their salts are the likely forms for solid phosphates deposited at high temperature in volcanic ash.

Heating sodium polyphosphates produced sodium trimetaphosphate:

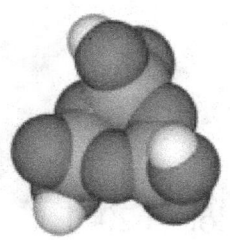

Author: Edgar181; Source Wikimedia Commons

Although biological systems can now readily utilize orthophosphate, orthophosphate (PO_4^{3-}) ions do not spontaneously form esters and amides.

$$[O_3P\text{-}O\text{-}PO_3]^{-4} + H_2O \rightarrow 2 \ PO_4^{3-} + 2 \ H^+ \qquad \Delta H \approx \text{-}23 \text{ kJ as written}$$

Thus, the pyrophosphates represent the major source of energy supporting living process until photosynthesis evolved.

The Current Heat Balance of the Earth

There is not much data on the early cooling of earth from a fluid metal core covered by liquid silicates and an atmosphere of molecular gases. But I believe that everyone agrees that the earth started off very hot and has cooled to its ambient temperature range. The initial heat presumably came from kinetic energy of crashing iron boulders. But the iron brought with it many radioactive isotopes, which decayed at various rates releasing thermal energy. It turns out that there are three elements (Uranium, Thorium and Potassium) with isotopes that have half-lives on the order of billions (10^9) years. Since these (depleted) elements still produce enough heat to keep the core of the earth molten, radioactivity was clearly a major driver of earth's temperature in the early stages of crustal formation.

Isotope	Half-life years	current %
^{235}U	4.468×10^9	99.2745%
^{238}U	0.7038×10^9	0.72%
^{40}K	1.28×10^9	0.012%
^{232}Th	14.05×10^9	99.98%

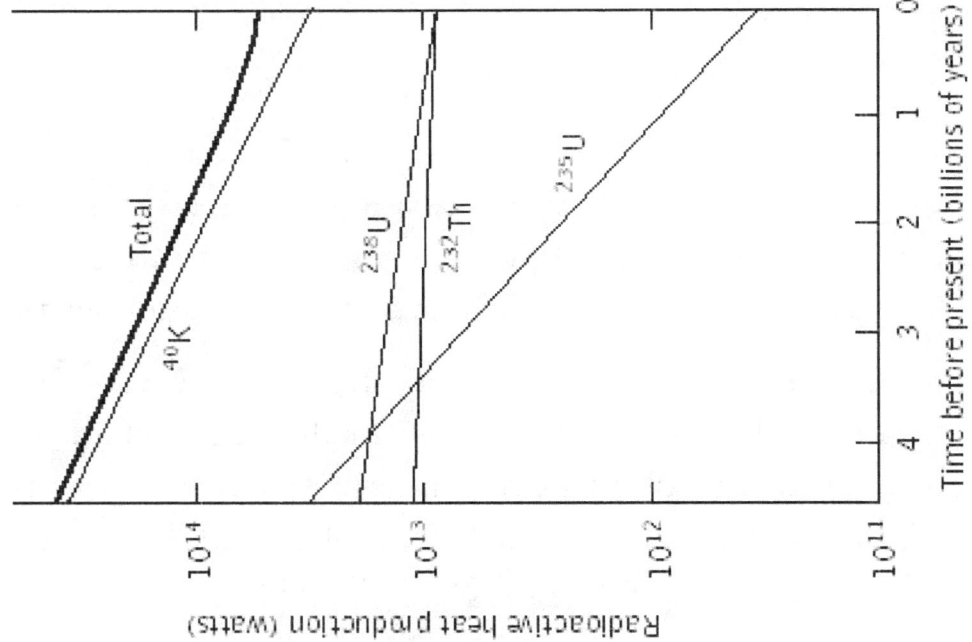

Source: http://science.jrank.org/pages/48045/radioactive-heat-production-in-Earth.html

Note that the scale on the decay graph is logarithmic and at the time of the consolidation of the earth's core roughly 10 to 20 times more heat was being released by these elements. The formation of planetary metallic cores probably coincided with reduction of the radioactivity from the short-lives (10^6 to 10^8 year) isotopes of various elements. But their energy is successfully radiated into empty space (which is nominally about $4°K$). Thus, the mantel developed a cooler crust of crystalline silicates and phosphates.

Lord Kelvin (1824-1907) calculated that the earth should have cooled to a solid mass within a few million years based on the heat of fusion of the molten core and mantel and the radiation of energy into space, but he was unaware of the tremendous amount of energy being released by radioactive decay. Today, only the very long-lived isotopes (particularly of thorium, uranium and potassium) remain. These isotopes still produce enough energy to keep the mantel molten (4×10^{13} Watts) and this will continue essentially constant for billions of years. The radiant heat from the surface of the earth to maintain this heat balance can be calculated as follows:

$$(4 \times 10^{13} \text{ W}) / (5.1 \times 10^{14} \text{ m}^2) = 0.078 \text{ W/m}^2$$

Today, only the top most layers of silicates have solidified (frozen). They are literally "flakes" of silicate crystals (dross) floating on a glassy viscous liquid mantel (slag). From the perspective of humans, we describe this movement as continental drift. The oldest rocks on earth are about 4.5 billion years old.

As the earth cooled further, water condensed (presumably first at the poles) from the atmosphere and settled into the low areas. In all likelihood, the water caused local cooling and contraction of the mantel and set up currents that moved the land masses. The crust became somewhat stable about 3.9 billion years ago with a thermal gradient in the crust of about $3.5°C/100$ m in depth.

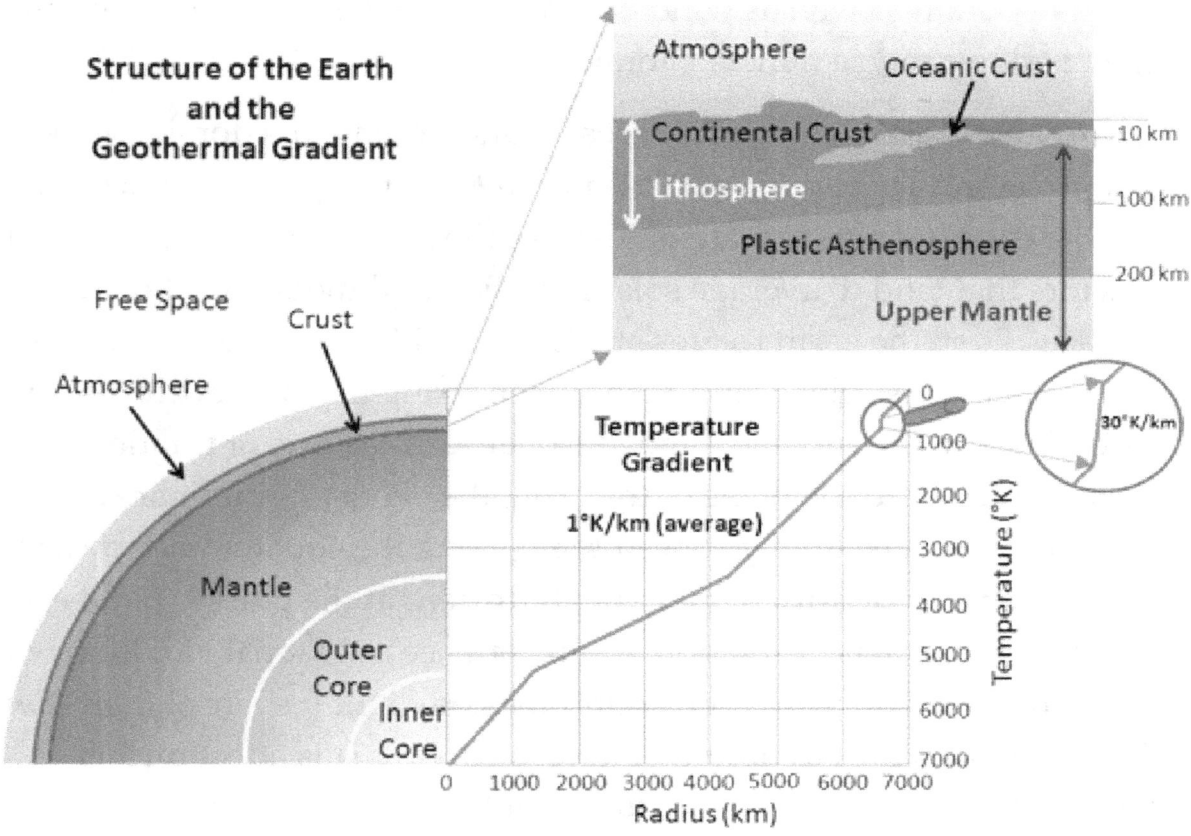

Source:

http://wattsupwiththat.com/2009/11/16/gore-has-no-clue-a-few-million-degrees-here-and-there-and-pretty-soon-were-talking-about-real-temperature/

After the decay of short half-life (<500 million years) isotopes and loss of heat from major collisions with other solar bodies (planetoids and meteorites)[24] the primordial earth began to cool.

Condensation: Atmosphere to Oceans

When the surface of the earth dropped below about 100°C (depending upon atmospheric pressure), liquid water could persist on the surface and the oceans

---

[24] The moon was produced through one of these collisions about 4.5 billion years ago.

began to fill up. Rain as well as wind began to erode the surface of igneous rocks and extract and hydrolyze minerals found there.

The sun provides radiant energy to the earth and became the dominant factor determining the surface temperature. Without the atmosphere and oceans, the surface of the earth would be more or less like the moon. Actually, the earth has several factors that tend to average solar radiation and moderate climate over its entire surface. First, the earth rotates at a fairly rapid rate on an axis slightly tilted to the plane of its orbit around the sun ensuring that all portions of the earth are exposed to the sun daily and seasonally. If the axis of rotation were exactly perpendicular to the plane of the orbit, the energy per unit area (watts/meter^2) would be highly concentrated along the equator and towards the poles the solar energy density would fall off (to zero at the poles). But, because of the slight tilt in the axis of rotation, we experience a seasonal variation with periods of continuous darkness and continuous daylight (i.e., insolation[25]) within the regions known as the Arctic and Antarctic circles. This effect further moderates the temperature over the surface of the planet.

The oceans and lakes have a large heat capacity (1 cal/g °C) and absorb solar radiation effectively. As water cools, it becomes denser and sinks, bringing warmer water to the surface. This would potentially result in complete freezing of the polar oceans top to bottom, but water (unlike virtually every other compound) expands when it freezes to ice, thus bringing ice to the surface where it is exposed to the atmosphere and insulates the liquid water below. The heat of fusion of water (80 cal/g) and the heat of vaporization of water (540 cal/g) also mitigate atmospheric temperature.

---

[25] *Insolation* is the incoming solar radiation. Not to be confused with "insulation," i.e., barriers to heat flow.

Global Cooling

The overall cooling trend (as radioactivity decayed) has been interrupted by the atmosphere. This has not been a smooth transition to an invariant homogeneity at least not at the scale of temperature variation of relevance to life. The chart below shows what we know about the last half-billion years.

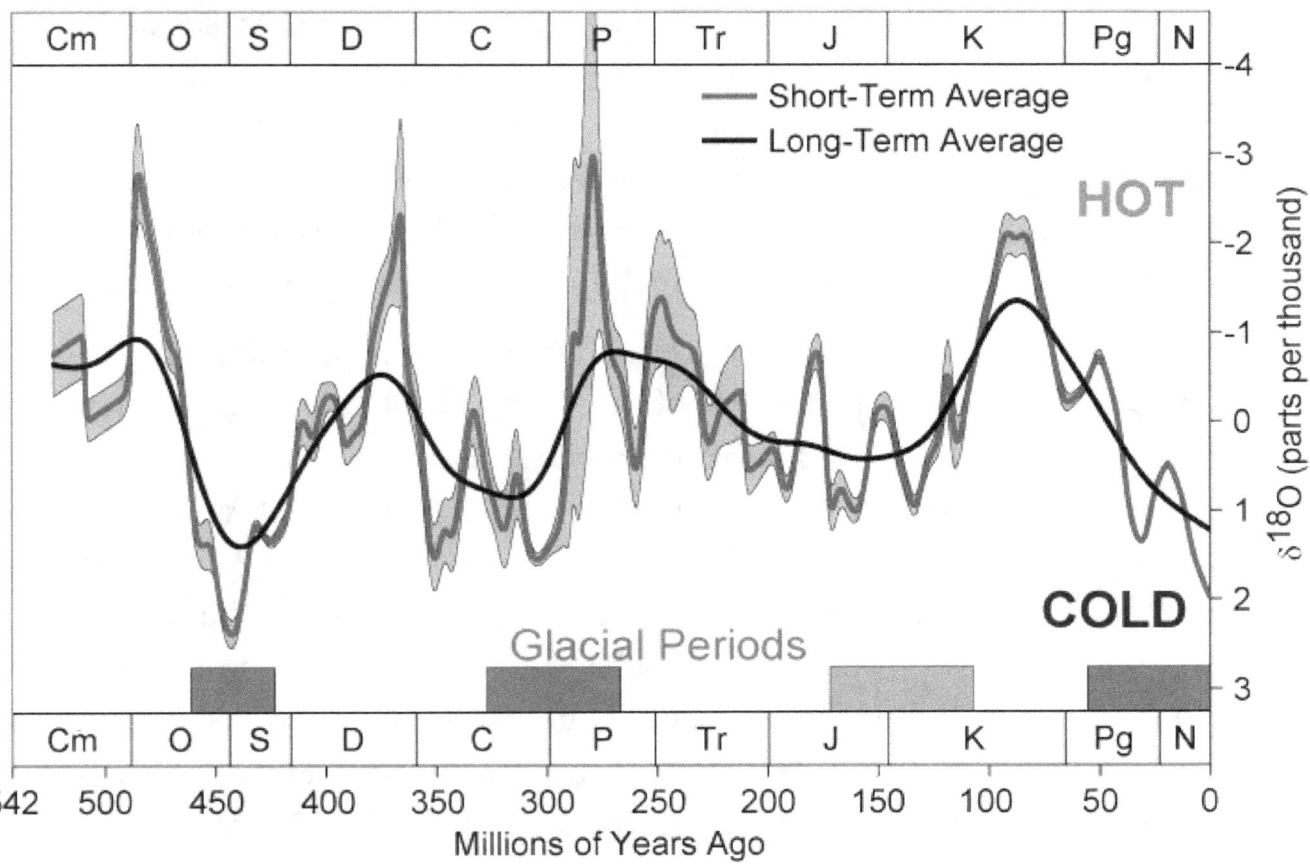

https://commons.wikimedia.org/wiki/File:Phanerozoic_Climate_Change.png

Author: Veizer et al. 2000; Source: Wikimedia Commons

In contrast to the smooth and predictable process of radioactive decay, there are a number of factors on the surface of the earth and in the atmosphere that impact heating on the interface of the crust and the lower atmosphere.

It must be noted that the earth cooled through periods where most of the water and carbon dioxide was in the atmosphere. In particular, water is universally recognized as the material in the atmosphere that has rescued the earth from decent into a barren ice-covered planet. The physics is basically this, referencing the solar spectrum (above) most of the solar energy reaching the earth is in photons with wavelengths between 350 and 700 nm. Most of it passes though the atmosphere (because the common gases [N_2, O_2, CO_2, H_2O] do not absorb much in this range) and impacts the solid surface where some is reflected and some is absorbed (as vibrations and rotations of molecules) and "degraded" into thermal motions. The surface of the earth, thus, has its own characteristic continuous radiation spectrum, but its peak is at much longer wavelengths than the sun.

The secret to earth's thermal stability is that water is a small, molecule with an inherently unstable rotation. Because of the strong dipole moment of water molecules, they absorb infrared light strongly in three bands (v_1, v_2, v_3 in the figure below) corresponding to molecular vibrations (e.g., $v_1 0 \rightarrow v_1 1$). There are also overtone absorptions corresponding to $v_1 0 \rightarrow v_1 2$, $v_1 0 \rightarrow v_1 3$, etc. at shorter wavelengths (i.e., less than 2000 nm).[26] Moreover, because water has a small moment of inertia, all of these vibrational bands are coupled with rotational energy level changes (i.e., when the bond lengths change, the moment of inertia must also change, $\Delta V +/- \Delta J$). The result is that water vapor absorbs strongly through most of the electromagnetic spectrum at longer than 10 micrometers = 10,000 nm).

Water molecules also have the same symmetry as wing nuts and constantly flip among axis of rotation (see Dzhanibekov Effect):

---

[26] The spectrum of light that is <u>reflected from within or transmitted through</u> water looks blue because of absorption of red light by overtones of the molecular vibrations. Of course, reflectance from <u>the surface</u> of water is similar to the solar spectrum.

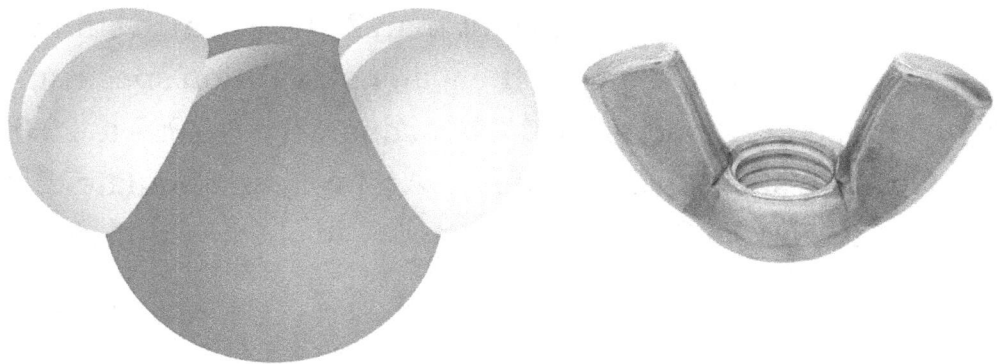

This allows the water molecule to absorb energy over long, continuous ranges of long wavelengths (greater than 20 microns). Thus, as long as the earth is cool (less than 220ºK) , the water vapor in the atmosphere effectively prevents *virtually all* radiation from the earth's surface into space.

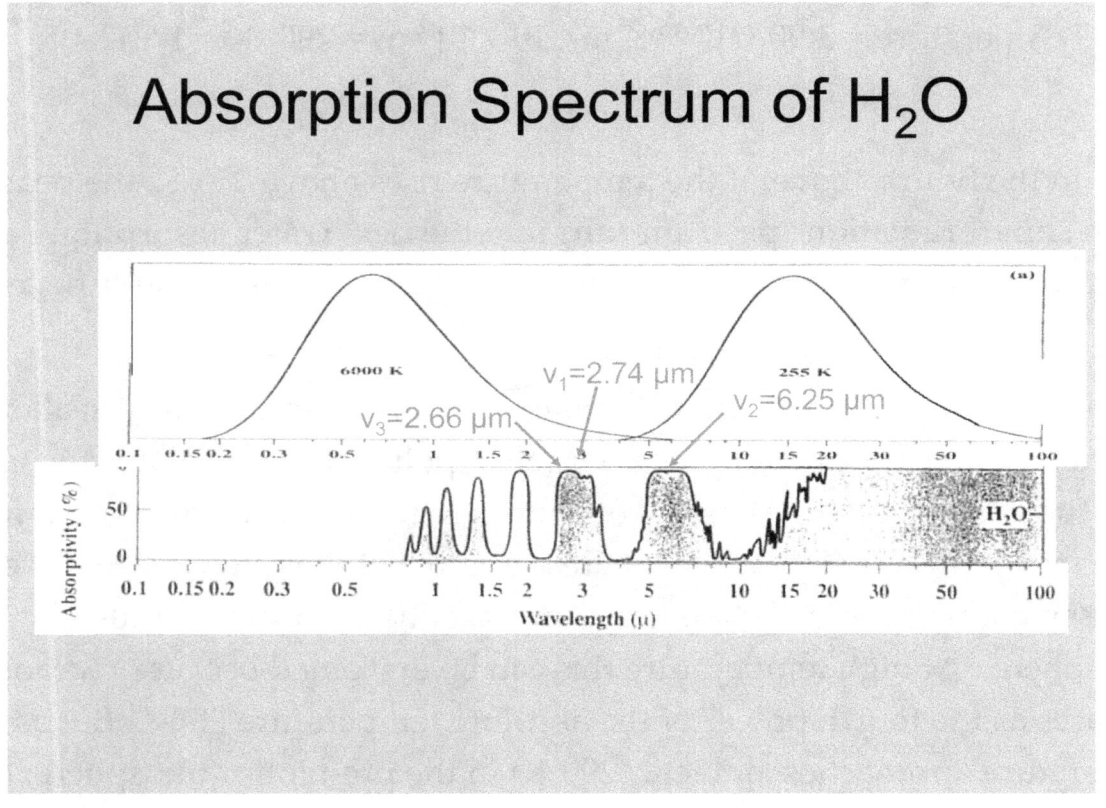

Source: http://ist-socrates.berkeley.edu/~budker/Physics138/Alyssa%20Atwood%20Atm%20Spec5.ppt#256,1, Atmospheric%20Spectroscopy

As the surface of the earth heats up (greater than 220°K), "earth shine" begins to escape to space through the first gap in the water absorption spectrum (between 6,000 and 20,000 nm; 6 to 20 microns μ). Thus, this edge (10,000 to 20,000 nm) tends to control earth's surface temperature. And we can calculate that temperature range from Wein's Law:

$$\text{Temperature} = 2.90 \times 10^{-3} \text{ °K m/ effective wavelength}$$

Which gives:

$$\text{Temperature} = 2.90 \times 10^{-3} \text{ °K m/ } 10 \times 10^{-6} \text{ m} = 290 \text{ °K} = 17 \text{ °C}$$

Note (in the figure above) if the temperature rises above 255°K, the peak of the "earth shine" radiation spectrum runs into the next water absorption band (7 to 10 microns). Thus, this window controls the temperature of the surface of the earth.

You probably have heard about "global warming" and the concern about carbon dioxide in the atmosphere. This is largely a political debate about the use of fossil fuels. The point that must be made is that the earth was cooling with most of the water vapor in that atmosphere and most of the CO_2 (which has ended up in massive limestone and chalk (calcium carbonate) formations also in the atmosphere. At high temperature this can be explained because radiant energy increases as the fourth power of the absolute temperature (T^4)[27]. But as the temperature approaches ambient (290°K) in the pre-biotic atmosphere, the CO_2 would still be highly elevated in the atmosphere because most limestone is of biotic origin and would not have formed as the earth cooled to 290°K. It turns

---

[27] Stefan-Boltzmann Law ($q = \sigma T^4 A$)

out that we know this is so because (i) calcium carbonate decomposes to CaO (lime) and CO_2 at moderate temperatures (see table below) and (ii) the limestone deposits on earth did not form until after life had evolved.

$$CaCO_3 \rightarrow CO_2 + CaO$$

Equilibrium pressure of CO_2 over $CaCO_3$

(P) versus temperature (T)

P (kPa)	0.055[28]	9.3	24	34	51	72	80	91	101	179
T (°C)	550	748	800	830	852	871	881	891	898	937

Source: Wikipedia "Calcium Carbonate"

There is more carbon tied up in limestone and chalk than in fossil hydrocarbons:

White Cliffs on the English Channel

The-chalk-cliffs-at-Dover.jpg (615×409) (tourismontheedge.com)

---

[28] The current pressure of CO_2 in the atmosphere is about 400 ppm or 0.0004 atm, 1 atm = 101 kPa; thus, the current concentration of CO_2 is 0.04 kPa.

The "global warming" alarm (which has been changed to the "climate change" alarm …which is still based on the assumption of global warming, which has not been proven), is tied to the fact that there is a CO_2 absorption band in the same range as the key water absorption edge (10 to 20 microns). But the absorption scales are logarithmic (not linear) …i.e., you cannot absorb *more than all* of the radiated "earth shine." Doubling the CO_2 concentration (in a spectral area where half the radiation is already absorbed by water vapor) does not have much net effect on absorption. Ironically, oxygen (O_2) has an absorption band in a much more critical spot (between 9 and 10 microns).

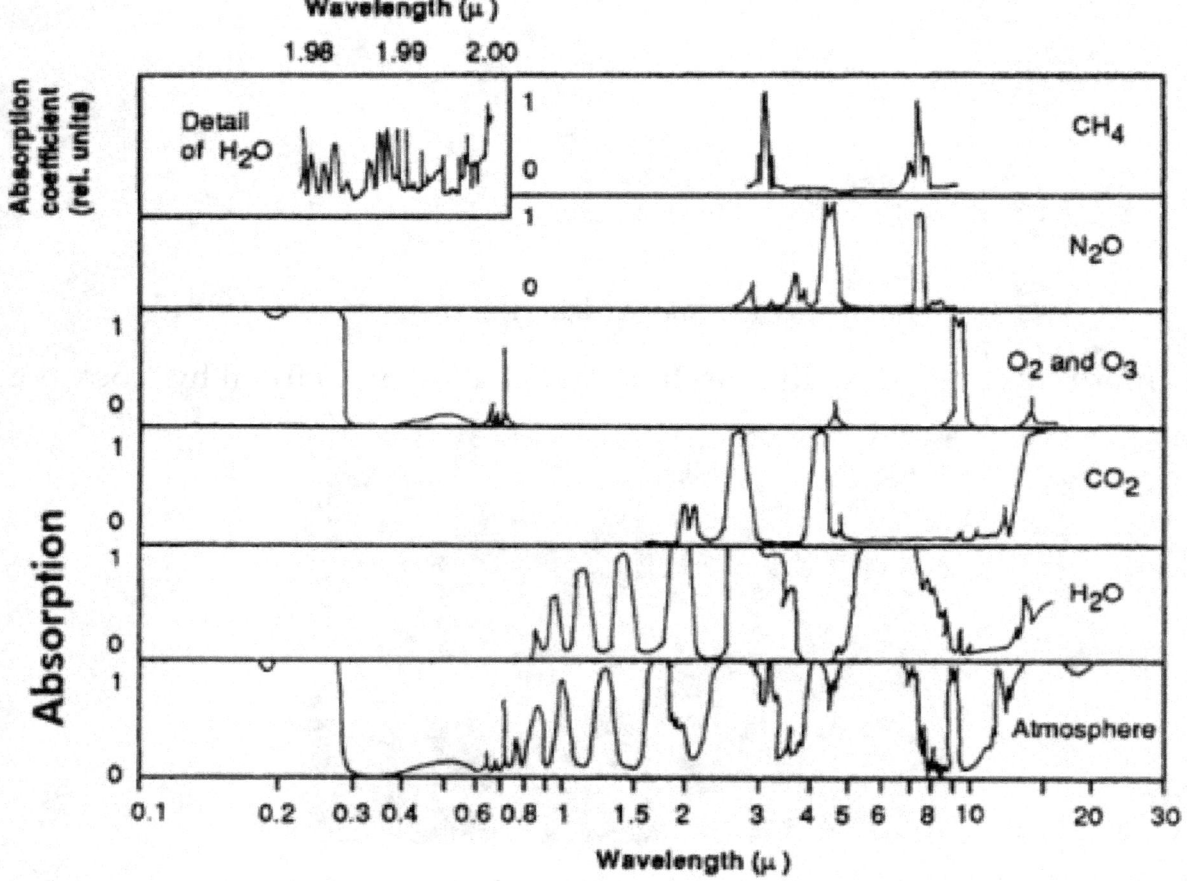

Figure 5-2. Absorption spectra for CH_4, NO_2, O_2, O_3, CO_2, and H_2O, and of the atmosphere. (From R. G. Fleagle and J. A. Businger [2006] after J. H. Howard [519] and R. M. Goody and G. D. Robinson [514])

Note that in the chart above, the authors have arbitrarily picked concentrations which saturate or nearly saturate (i.e., max-out) the spectrum. They arbitrarily chose concentrations of water vapor and carbon dioxide in which the CO_2 contribution of absorption was maxed below about 15 microns and the water spectrum is not. In the "atmosphere" spectrum (where CO_2 is a small contributor) we see that the water spectrum is the major contributor to absorption in this range. (And of course, liquid water is maximizing absorption of incoming solar radiation in the infrared region 0.8 to 2.0 microns.)

With respect to combustion of coal (C) to produce CO_2 it seems relevant that for every molecule of CO_2 formed one molecule of oxygen is consumed.

$$C + O_2 \rightarrow CO_2$$

Reducing O_2 content would seem to be as important as reducing CO_2 content. Of course, because of the logarithmic effects, combustion of coal has little effect on the absorption of "earth shine" by oxygen.

But absorbing outgoing "earth shine" is only one element for our temperature balance. Ironically, the thermal problem facing life is not runaway heating (which appears to have never occurred) but rather runaway cooling (Look up "Snowball Earth").

Surface	Albedo
Fresh Snow	0.6-0.9
Cloud Tops	0.4-0.8
Old Snow	0.4-0.6
Dry Sand	0.4
Wet Sand, Dry Soil, Deserts	0.2-0.3
Crops and Meadows	0.1-0.2
Dark Wet Soil, Forest	0.05 to 0.15
Liquid water	0.08

 Basically, when water condenses into clouds or precipitates as snow, the brilliant white surface that is created has a very high albedo[29] (these surfaces reflect most of the sunlight before it is ever absorbed by the earth or lower atmosphere).

Source: https://i.ytimg.com/vi/OosKvX1-BPs/maxresdefault.jpg

Ice reflects much more solar radiation than water (i.e., the albedo of snow and sea ice is about 0.65 while the albedo of ocean water is only 0.07).[30] Thus, when ice forms over large areas of the planet, the planet tends toward cooling. This effect is magnified because at lower temperature, the absolute humidity (i.e., amount of water vapor in the atmosphere) also decreases. These effects seem to periodically (within the history of humans) bring on ice ages (i.e., polar ice advancing into middle latitudes) and on a couple of occasions have almost

------

[29] albedo is the fraction of light reflected from a surface

[30] J A Coakley 2003. Reflectance and Albedo. Oregon State University, Corvallis, OR, USA Copyright 2003 Elsevier Science Ltd.
http://curry.eas.gatech.edu/Courses/6140/ency/Chapter9/Ency_Atmos/Reflectance_Albedo_Surface.pdf

completely covered the planet with ice (or slush). These "snowball planet" episodes probably ended when major volcanic eruptions or meteoric impacts covered the snow with light-absorbing ash and returned water and carbon dioxide to the atmosphere. It is relevant that there is no evidence that all the oceans have ever evaporated since they formed (nonetheless, the Mediterranean Sea did evaporate completely).

The negative feedback mechanisms preventing over heating (such as near complete reflectance of sunlight from the tops of clouds formed when water evaporates) seem to ensure that the planet cannot overheat. It seems likely that the mass of earth's atmosphere was determined by progressive loss of water, nitrogen, etc. until a thermal balance that could not produce runaway heating was achieved. If the atmosphere were much more massive (i.e., much more nitrogen gas) the atmospheric pressure at the surface of the oceans would be higher and the rate of water evaporation would be reduced at any given temperature. Thus, the temperature would increase causing loss of nitrogen to space. The point is that the amount of earth's atmosphere is not an accident, but rather a system determined by setting an upper limit on the temperature that the atmosphere can achieve. Note that water vapor (H_2O molecular weight 18) should escape faster than nitrogen (N_2, 28) from the atmosphere. Recall that nitrogen was the primary gas in the atmosphere before photosynthesis. Importantly, the initial moles of *nitrogen gas* when the planet formed should be similar to the *total moles of combined and free oxygen* today, based on the CNO cycle for nuclear fusion.

The evidence for snowball earth had accumulated in the 1800s, but did not make sense until the theory of plate tectonics explained where the various land masses were located when certain glacial deposits were laid down. This all came together in 1964 under Walter Brian Harland (1917–2003).

Source: https://i.ytimg.com/vi/P8q13Fqj6MA/maxresdefault.jpg

2.0 Chemical and Physical Processes leading Living systems

Genesis 1(20): *And God said, Let the waters bring forth abundantly the moving creature that hath life, and fowl that may fly above the earth in the open firmament of heaven.*

Charles Darwin letter to Joseph Hooker (February 1, 1871): But if (and oh what a big if) we could conceive in some warm little pond with all sorts of ammonia and phosphoric salts, — light, heat, electricity &c. present, that a protein compound was chemically formed, ready to undergo still more complex changes, at the present day such matter wd be instantly devoured, or absorbed, which would not have been the case before living creatures were formed.

2.1 The Primeval Soup Theory

Most religions and especially those monotheistic religions traced to Moses strongly relied on Devine intervention for the origin of the universe and life. The mechanism was of no importance to believers, but I would point out that the scripture does not require that there be no mechanism following from natural laws. But, somehow the creation of life (as well as the physical universe) was considered to be off limits to scientific analysis well into the 1800s as proven by the anxiety of Charles Darwin to publish his theories of evolution.

There had long persisted a theory of *vital force* involving the direct intervention of God to accomplish otherwise impossible transformation. Aristotle (On the History of Animals, Book V, Part 1) had proposed spontaneous generation:

> *So with animals, some spring from parent animals according to their kind, whilst others grow spontaneously and not from kindred stock;* ***and of these instances of spontaneous generation some come from putrefying earth or vegetable matter, as is the case with a number of insects, while others are spontaneously generated in the inside of animals out of the secretions of their several organs****.*

But by 1668, the devout Francesco Redi disproved at least some cases of spontaneous generation and argued that *life comes only from life.*

This was the doctrine followed by Charles Darwin, but other biologists of the 1800s were willing to consider a chemical origin of life. Even strong supporters of Darwin's theory of natural selection (1859) viewed his reluctance to extrapolate backward to an inorganic to organic (living) transition as a mistake.

Bacteria

The invention of the microscope[31] and the discovery of bacteria (circa 1870), however, led to a school of thought that they were the missing link from inorganic materials to living tissues. The result was a variety of claims of spontaneous generation (abiogenesis), which were ultimately debunked by Louis Pasture (1859). This result actually strengthened the support for Devine intervention. If bacteria could not self-assemble from their constituents (after dissolution by boiling) how could inorganic materials ever give rise to life without help from God? For example, a paper in 1876 by William Roberts[32] is essentially a rebuttal to claims by a Dr. Bastian that he had successfully demonstrated spontaneous generation from various (organic) media. Roberts points out that denial of the spontaneous regeneration of bacteria does not eliminate the possibility of abiotic origins of life:

> *The reluctance of some evolutionists to give up the spontaneous origin of bacteria is evidently due to the notion that this question is bound up with that of abiogenesis generally. This is a wholly erroneous idea.* **The question of abiogenesis will still remain after all have acquiesced in Pasteur's views of the origin of bacteria: indeed, to a logical evolutionist there would appear to be a strong a priori improbability in the abiogenic origin of bacteria. They were not wanted, and could not exist, on the earth's surface until after other organisms had lived and died before them.** *Their special function and feeding ground lie amid the wreck of living things. And if the survival of the fittest hold good in regard to bacteria,* **they must be the remote progeny of less perfect organisms of the same class.** *What can be more perfect than their adaptability to their place and use in the order of Nature? They resist, in certain media, for considerable periods the heat of boiling water; they multiply with*

---

[31] Antony van Leeuwenhoek invented the microscope and discovery of protozoa (animalcules) dates to about 1676. But bacteria were much smaller and more primitive.

[32] Roberts W. 1876. A Word on the Origin of Bacteria, and on Abiogenesis. *Br Med J.* 1(792):282-3.

*incredible speed; their germs survive in countless myriads in the dust of the atmosphere; they float in every drop of water on land and sea; they appear to be omnipresent and almost indestructible. **Those who are in search of a case of abiogenesis, should seek among the primitive organisms-if there be any such-which can exist and grow amid inorganic elements,** in the water of the sea, or the mineralized springs and streams of the land. When Pasteur says that abiogenesis is a chimera, he prudently adds, "in the present state of science"; and even thus qualified, the expression is perhaps too strong. But it is absolutely certain that up to the present time no case of abiogenesis has been presented which has stood the test of accurate investigation; nor can it be doubted that, in so far as the antiseptic treatment of disease rests on the origin of bacteria, the advocates of that treatment stand on unassailable ground.*

Clearly, Roberts assumed that contemporary bacteria only existed as a parasite feeding off the remains of more evolved life forms. But he sees them as specialized descendent of more primitive organisms of the same type.

By 1900, chemistry was well enough established for biologists to realize that "organic" chemicals could arise from inorganic sources. Indeed, this had been shown in 1828 when Friedrich Wohler (1800-1882) heated ammonium cyanate and produced urea. Urea was clearly known to be an "organic" material; and according to the prevailing theory of *vital force*, urea was presumed to only be formed by living organisms. This breach of the barrier between the inorganic world and the organic world, cracked open the door for an abiotic origin of life (contrary to the prevailing interpretation of the scriptures.

Indeed, the leading advocate of the theory of *vital force* was the highly regarded chemist Jacob Berzelius (1779-1848). After his death (1848) *vital force* fell away among the chemists, although it was still realized that organisms produced many materials not readily accessible by non-biological means. Most of these chemicals contained carbon and chemistry is still (2021) divided between organic (carbon-based) and inorganic disciplines.

The Hypotheses of Oparin and Haldane

Alexander Ivanovich Oparin (1894–1980) and John Burdon Sanderson (J.B.S.) Haldane (1892–1964) are identified as the immediate inspiration for the first experimental investigation of the origins of life on earth.

From the time of Darwin (mid-1800s), biology had been focused on the evolution of life and speciation. This fascination continues today. But the origin of life was conceded to religious thought by almost every one including Darwin. Perhaps the idea that began to shift focus was *panspermia,* which became popular in the late 1800s.[33] With growing understanding of the universe, and extrapolation of physics and chemistry know on earth to the universe, the idea that there might be life and civilization on other planets and in other solar systems began to come into focus.

Indeed, life on Mars was very seriously considered. Based on the apparent presence of water on Mars and apparent changes in the color of Mars associated with seasons, Italian astronomer Giovanni Virginio Schiaparelli (1835–1910) published a map of Mars (1877), which was widely interpreted as suggesting active rivers and seas:

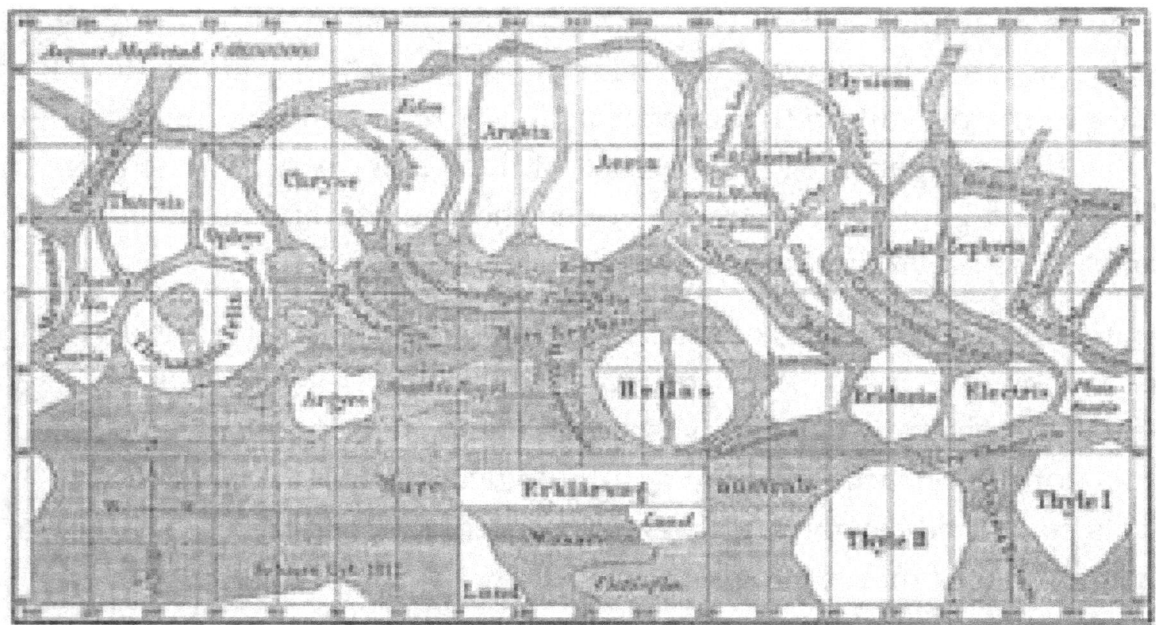

Source: https://en.wikipedia.org/wiki/Giovanni_Schiaparelli

---

[33]Ludwig von Helmholtz (1821–1894) and William Thomson, 1st Baron Kelvin (1824– 1907).

In 1889, new channels were observed and "confirmed" by American astronomer Charles A. Young. It was not a big jump from the idea of *natural channels* for flow of water on Mars to *intelligently constructed canals*. Percival Lawrence Lowell (1855–1916) popularized and disseminated the idea that Mars was covered with a series of canals carrying water from the polar caps to the dry lower latitudes of Mars. For at least the next 50 years, speculation about Mars in the popular media varied from the dangerous invaders envisioned in *War of the Worlds* (H. G. Wells, 1897) to the funny Warner Brothers' (Looney Toons) cartoon characters.

https://i0.wp.com/bloody-disgusting.com/wp-content/uploads/2018/07/War-of-the-Worlds.jpg?resize=785%2C442&ssl=1

https://i.stack.imgur.com/mpLK3.jpg

It was actually a fun time that inspired men like Werner von Braun (1912-1977) to explore other worlds.

Alexander Ivanovich Oparin (1894–1980) was a chemist who focused on reactions catalyzed by plant enzymes. With the atheistic Communist takeover of Russia in 1917, it became easy (actually virtually required) for Oparin to adopt a god-less view of all biological processes first published in 1938 as *The Origin of*

Life.[34]   It is important for what has followed that the presence of methane (CH_4) in great quantities had just been reported in the atmosphere of Jupiter. Indeed, Jupiter has an abundance of light (H_2, He) and the hydrogen-rich gases (methane, ammonia, hydrogen sulfide, water, phosphine and arsine). Ignoring the possibility that these gases had been stripped from the inner planets (see above), it was postulated that the primitive atmosphere of all planets had been dominated by these gases. The modern understanding of the atmosphere of Jupiter is presented below.

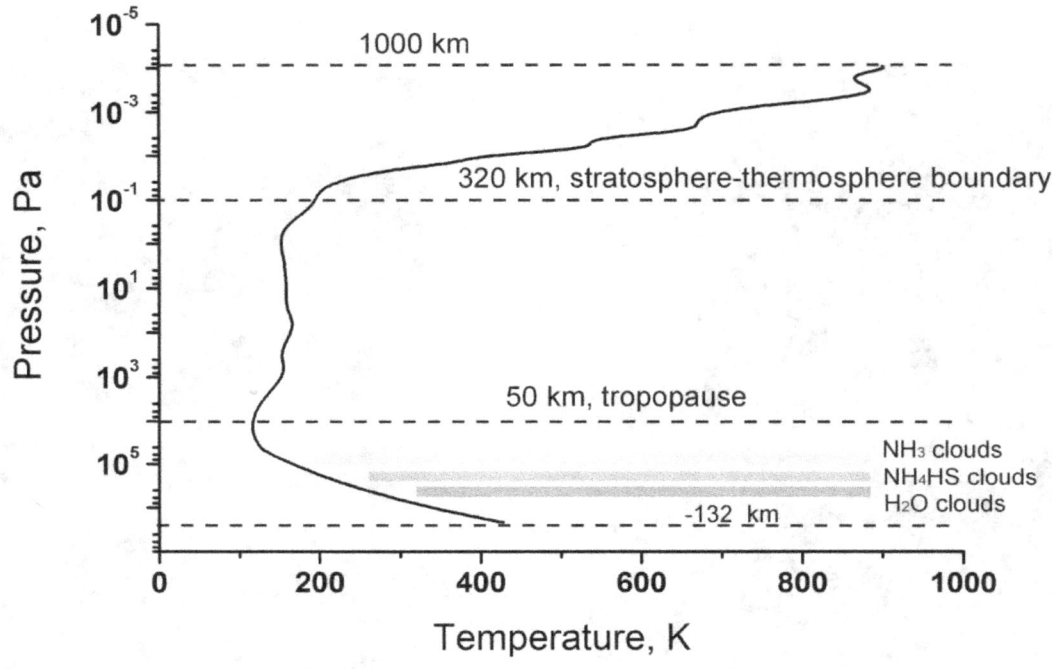

Source:
https://upload.wikimedia.org/wikipedia/commons/f/f5/Structure_of_Jovian_atmosphere.png

Starting with this conception of the primordial atmosphere of earth, Oparin speculated (1924) that life was gradually formulated in the oceans.

---

[34] Oparin, A.I. The Origin of Life. Moscow: Moscow. Worker publisher, 1924 (in Russian) (first English translation published in 1938). English translation: Oparin, A.I. The Origin of Life. New York: Dover (1952)

J.B.S. Haldane (1892-1964) was British and served in the British army in WWI (1914-18); but soon after that, he was drawn to the idealism of Marxism. In 1924 he published *Daedalus* in which he anticipates (rather cynically and pessimistically) many of the possibilities that biology and medicine have opened for humanity…primarily human control of evolution and personal characteristics. Obviously, Haldane would have been anti-capitalistic and familiar with and sympathetic to the views of Oparin on abiogenesis. In 1929, Haldane published "The Origin of Life," in the *Rationalist*. Basically, all he did was concur that water, carbon dioxide and ammonia (in the oceans) contained all the elements (C, H, N, O) needed to make the most critical biomolecules. Which formed viruses. Eventually an "oily film" allowed the encapsulation of components into cells.

Frankly, from a chemical standpoint I would expect this much from any high school biology student.

2.2 The Urey-Miller Experiments

Harold Clayton Urey (1893–1981) obtained a PhD in chemistry in 1923 and became a professor of chemistry in 1931. The neutron was discovered in 1931 and Urey soon explained the isotopes of hydrogen (^{1}H, ^{2}H (deuterium) and ^{3}H (tritium)) in terms of protons and neutrons. Obviously, Urey was not particularly interested in abiogenesis. However, he had a graduate student (Stanley Miller, 1930–2007) who was.

Following, the speculation about the early atmosphere of earth, Miller assumed that the atmosphere consisted of ammonia, methane, and hydrogen. Miller was allowed to investigate if elementary biomolecules (such as amino acids) could be formed from these compounds under the influence of a high energy source. Electrical discharge (simulating lightning) was chosen as the source of energy. What happened next sounds a little "dodgy." Miller did the classic and widely known experiment. And used simple thin layer chromatography to "identify" traces of two amino acids.

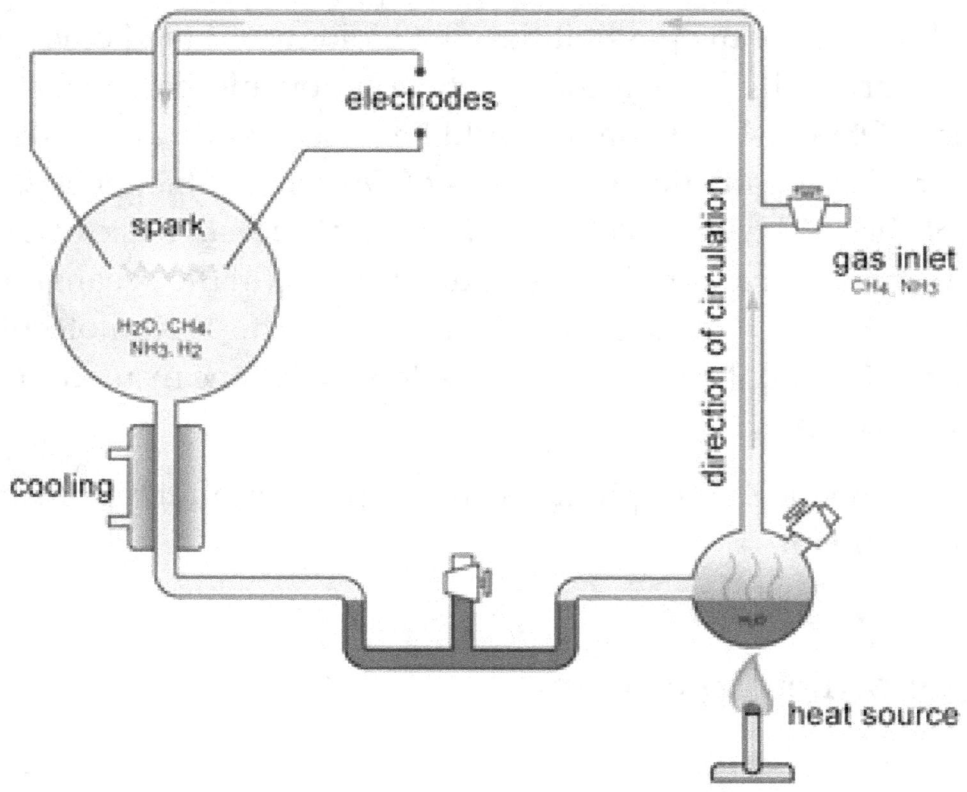

Author: Carny at Hebrew Wikipedia; Source: Wikimedia Commons

If you are familiar with chromatography in the early 1950s and both its qualitative and quantitative limitations, you will appreciate that this was not exactly a clear-cut result. Indeed, Urey actually did not want anything to do with the paper and had Miller submit it to Science with only his name on it. Science held the paper for quite a while and only agreed to publish with Urey's name when he threatened to take the paper to the *Journal of the American Chemical Society*. Their paper was published in *Science* (15 May 1953) and has become a fundamental cornerstone in biology because they claimed to isolating traces of amino acids. This, of course, made Miller's career and over the remainder of his life he modified the conditions of the experiment as did others. With the improvements in chemical detection and characterization, this approach has been successful in isolating traces of many amino acids and other organic molecules associated with biology.

Unfortunately, I am not impressed. I certainly am skeptical of the original claims and the fact that Miller and his graduate students seem to be the only ones following up this work is a red-flag for me. *Science* has been known to publish a few dubious papers.

The amount of energy involved in these experiments simply shreds molecules and allows the gradual accumulation of every stable combination and I would seriously doubt that amino acids would be produced in detectable quantities.

2.3 The Carbohydrate-Based Hypothesis

Pardon me for not building upon what I consider to be the dead-end experiments of (Urey and) Miller. I think they were seriously misled regarding the composition of the primordial atmosphere of the earth and selected the wrong source of energy.

The Primordial Atmosphere of Earth

I spend the first part of this book explaining how the earth's atmosphere was formed. The major components at the time that the curst cooled to the point that complex molecules would be stable (about 400°C). The vapor pressure of water at 300°C is upwards of 100 atmospheres (1000 psi). The critical point of water is 373°C and 220 bars, i.e., at this temperature the kinetic energy is greater than the intermolecular attractions and the molecules can only be held together (in contact) by substantial pressure. I believe most amino acids and sugars would decompose over 300°C via decarboxylation[35] or deamination. At this time, the atmosphere was primarily N_2, CO_2, H_2O, NH_3 and significant amounts of

---

[35] Douglas M. Jackson, Robert L. Ashley, Callan B. Brownfield, Daniel R. Morrison & Richard W. Morrison (2015) Rapid Conventional and Microwave-Assisted Decarboxylation of L-Histidine and Other Amino Acids via Organocatalysis with R-Carvone Under Superheated Conditions, *Synthetic Communications*, 45:23, 2691-2700, DOI: 10.1080/00397911.2015.1100745

formaldehyde $H_2C=O$. Most certainly no hydrogen or methane was present. Complex organic molecules likely did not begin collecting on the mineral crust surface of the earth until it dropped below 200°C.

It is likely that clouds of water vapor formed a "false surface" of the earth upwards of 20,000 feet (5 km) above the surface. In the atmosphere above this cloud layer, sunlight (very much as today) penetrated into an atmosphere that was denser than experienced on the surface of the earth today. Water-soluble compounds generated in this zone above the clouds would be accumulated in the water droplets of the clouds at about 100°C. A perpetual cycle of rain and evaporation cooled the earth as the radioactivity of the core died away. Light from the sun was not a significant factor in warming the earth because most of it was reflected off of the cloud tops.

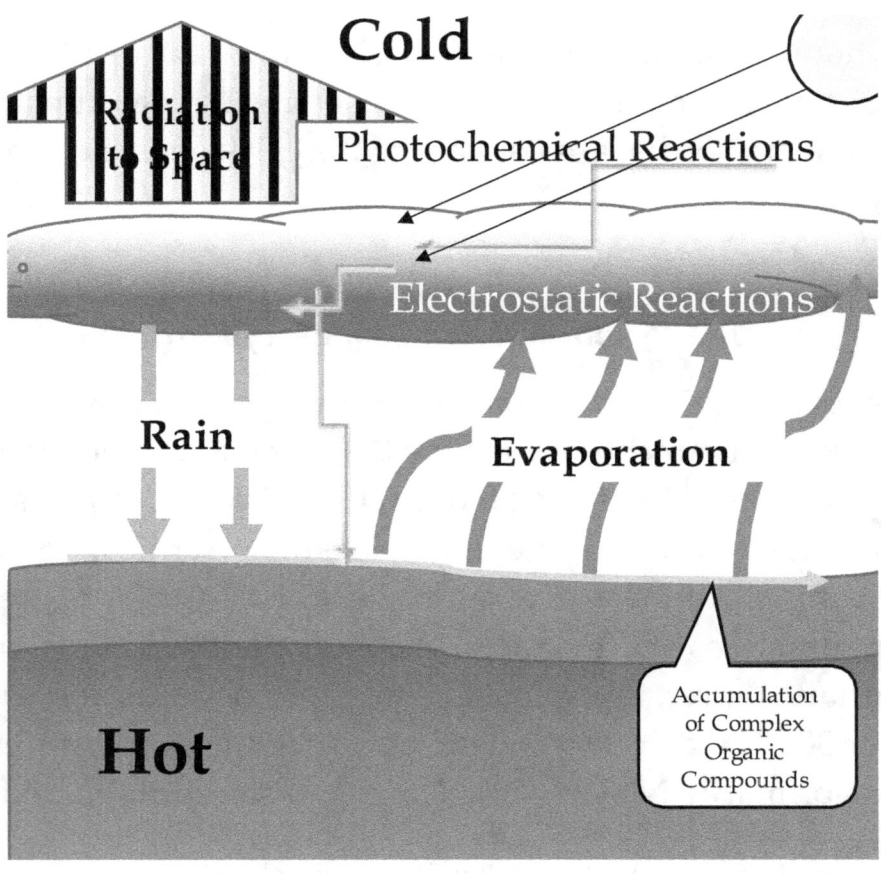

The Basics of Photochemistry

I am eagerly hoping and many biologists are reading this book, and thus, I am including some general material such as the following.

The photon energy of monochromatic light is determined by the Planck's equation:

$$E_{photon} = h\nu$$

where h is Planck's constant; ν is frequency; and since the relationship between frequency (ν) and wavelength (λ) can be calculated using the speed of light (c),

$$\nu = c/\lambda$$

we can write the equation as follows:

$$E_{photon} = h\nu = (hc)(1/\lambda)$$

Instantaneous conservation of energy requires that absorption of a quantum of energy is an all-or-nothing event. This is a very important principle because what it means is that in order for matter to absorb a photon of light (and acquire the energy that the light carried), two conditions must be met: (i) There must be a physical mechanism for the matter to interact with the light wave and (ii) the physical system must have an energy transition that is exactly equal to the photon energy. The first criterion is generally met by polar molecules[36] (i.e., molecules that have electric or magnetic dipoles due to asymmetrical distribution of electrical charge [+/-] or a magnetic dipole) that can interact with the oscillating magnetic and electric fields that are perpendicular to the direction of travel of the ray of light. The second criterion may be met by any of the electro-mechanical systems of a molecule including: nuclear, electronic, or kinetic (vibrational, rotational and translational) energy states.

---

[36] In most cases, I use the term *molecule* to refer to any collection of bonded atoms including neutral molecules, free radicals (i.e., neutral fragments of molecules) and ions (i.e., charged fragments of molecules + or -). Cation refers to a [+] charge and anion refers to a [-] charge.

We typically associate certain types of energy systems with the type of radiation that that interact with (i.e., absorb and emit photons):

nuclear transitions   gamma rays   (<0.1 nm)[37]

electronic transitions

>*inner core* (k-shell) x-rays (0.1-10 nm)

>*valence shell* (sigma bonds)   ultra-violet (UV) (10-300 nm)

>**delocalized** (pi bonds and d orbitals) visible (300-800 nm)

Mechanical transitions

>*Vibrations* infra-red (IR) (800-20,000 nm)

>*Rotations* microwaves (20,000-1,000,000 nm)

>*Translations* radio waves (>1,000,000 nm)

If you look at the solar spectrum (6000°K blackbody radiation), you notice that most of the light reaching the earth is in what we now call the "visible spectrum" centered around 500 nm wavelength, but also including a significant amount of UV light down to 100 nm.   The spectrum received from the sun (blackbody radiation) covers a continuum of wavelengths, thus, sunlight can match up with almost any important electronic transition of complex molecules.

---

[37] Gamma rays and x-rays were fairly rare after the earth has formed. However, high-kinetic-energy charged particles (protons, electrons) continue to impact the upper atmosphere and may be instrumental in initiating condensation and cloud formation (like in a cloud chamber).

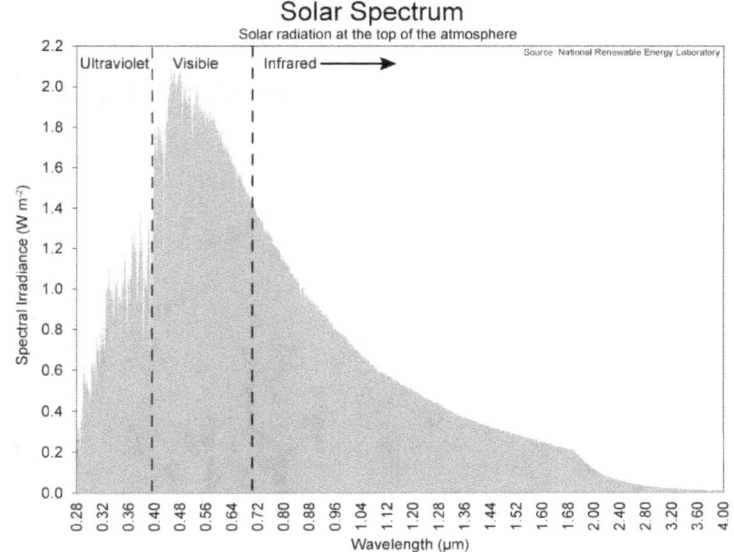

Source: http://sites.gsu.edu/geog1112/solar-radiation-seasons/

It can be shown from Planck's law and knowledge of the sigma bond energies of molecules that UV light with wavelengths shorter than about 300 nm are capable of breaking C-C, C-H, C-N, C-O single chemical bonds.

$$X\text{-}X \rightarrow h\nu \rightarrow X\cdot + X\cdot$$

It can be deduced that if the sun were much hotter or much cooler (regardless of how far we were from it) that life could not have evolved on earth because there would not have been proper wavelengths of light to produce reaction in the atmosphere (cooler sun) or the light would have destroyed the compounds that were produced (hotter sun).

In addition to breaking sigma bonds (single bonds directly between two nuclei), less energetic photons can be absorbed and activate the pi bonds associated with double bonds:

$$X\text{=}X \rightarrow h\nu \rightarrow X\cdot - X\cdot$$

In particular, activation of double bonds (pi bonds) can lead to concerted (i.e., "happening all in one step") cyclization reactions (2+2 cycloaddition):

X=X + X=X →hv→ four-membered ring (X₄)

An example:

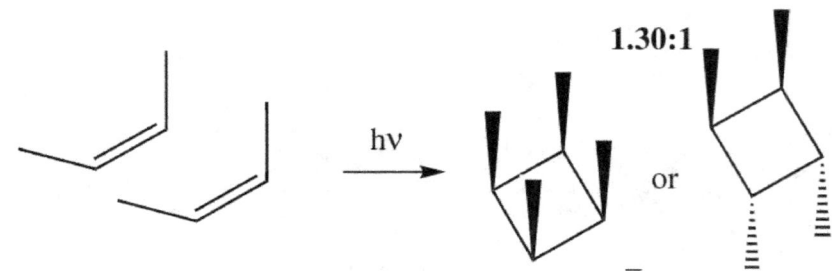

Source: https://www.ch.ic.ac.uk/local/organic/pericyclic/p38.svg

Chemical Reactions Above the Cloud Layer

The presence of formaldehyde and ammonia in the atmosphere above the cloud layer must lead to an equilibrium between the aldehyde and the imine:

$$+ NH_3 \leftrightharpoons H_2C=NH + OH_2$$

Actually, the reaction can continue all the way to hexamethylene tetraamine (e.g., hexamine) if water is removed:

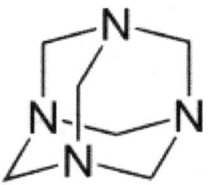

Thus, there are six possible combinations for 2+2 cycloadditions in the early atmosphere exposed to sunlight above the clouds:

O=C=O + O=C=O

$H_2C=O$ + $H_2C=O$

$H_2C=NH$ + $H_2C=NH$

O=C=O + $H_2C=O$

$O=C=O \ + \ H_2C=NH$

$H_2C=O \ + \ H_2C=NH$

And if we focus on only those cases where stable C-C bonds are formed (the other cases are probably easily hydrolyzed back to the starting materials by addition of H_2O) the other members of the four member rings will included either a O-O or O-N bond. Which are not very stable and are easily broken by sunlight (hv) leading to rearrangements of the ring molecules:

$$H_2C=O + H_2C=O + hv \rightarrow \overset{\displaystyle O-O}{\underset{\displaystyle H_2C-CH_2}{| \quad |}} \rightarrow O=CH\text{-}H_2COH$$

Key intermediate to carbohydrates

$$H_2C=NH + H_2C=O + hv \rightarrow \overset{\displaystyle O-NH}{\underset{\displaystyle H_2C-CH_2}{| \quad |}} \rightarrow HN=CH\text{-}H_2COH \ +$$

$O=CH\text{-}H_2CNH_2$

Key intermediates to glucosamines

Note that the combination with CO2 reacting with any of the other compounds would produce a four-membered ring (nominally 90 bond angle) with an external C=O which requires a bond angle in the ring of 120 degrees. The strain in such a ring precludes it from being isolated (i.e., if they form, they decompose).

Glucose and Fructose

You will notice that in each case the rearranged product has either an RCH=O or RCH=NH element. There is nothing stopping these compounds from continuing to add more $H_2C=O$ or $H_2C=NH$. Since $H_2C=O$ is much more prevalent in the atmosphere than $H_2C=NH$, these subsequent additions, lead directly to sugar molecules (carbohydrates).

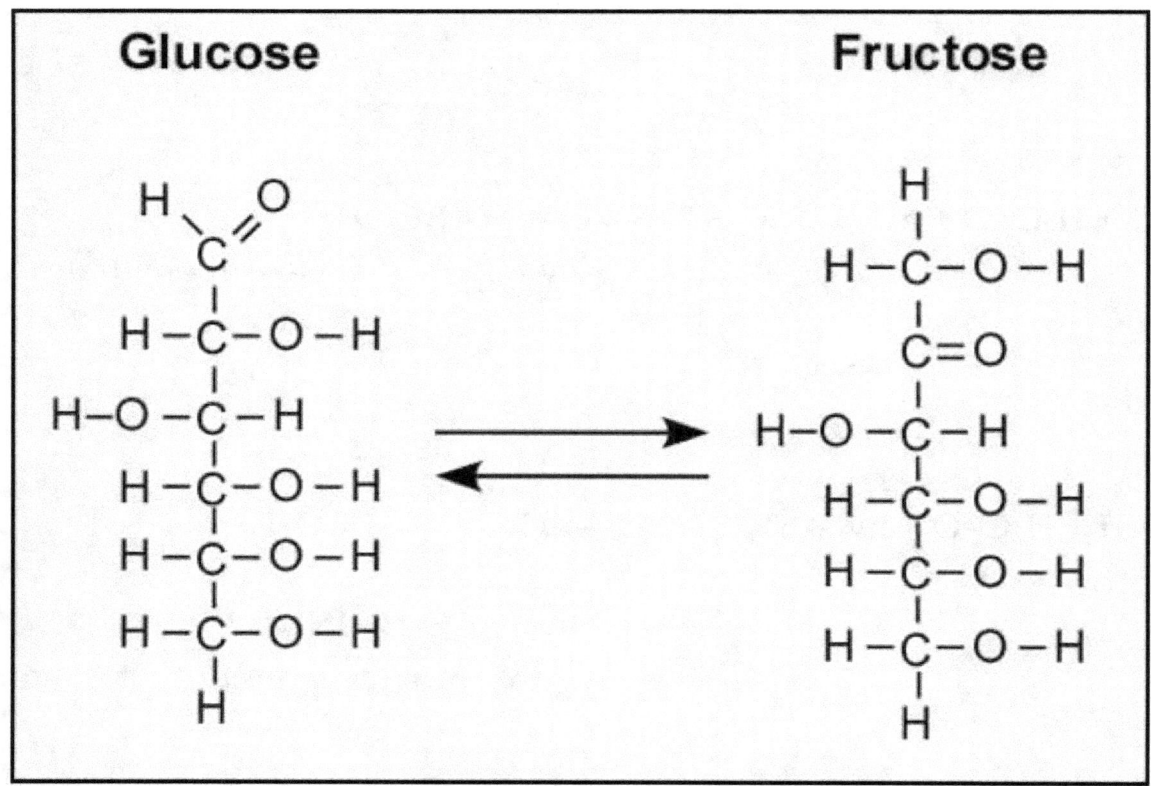

Source: http://4.bp.blogspot.com/-
CuW0eRzf32Q/VUZORpsHRmI/AAAAAAAACDA/2IiVL5wuPv0/s1600/glucose-
fructose1.jpg

You will note that when the aldehyde is formaldehyde, R = H and we are forming the backbone of carbohydrates. It is relevant and important that when the carbon chain reaches a length of 5 or 6 units, the polymerization is likely to be interrupted by formation of stable 5- and 6-membered (hemiacetal) ring compounds. The bond angle around sp3-hybridized carbon are 109 degrees and the six-membered ring allows a cyclic structure with no ring strain and formation of a hemi-acetal:

$$\text{ROH + RCHO} \leftrightarrows \text{RO-CRH-OH (Hemiacetal)}$$

Since this is formed within the same molecule, there is little unfavorable entropy and the formation of a C-O sigma-bond (~ -358 kJ/mole) is favored over a C-O pi-bond in an aldehyde (~-322 kJ/mole) by 35kJ/mole. When a total of 6 units of formaldehyde are thus polymerized, the molecules are capable of forming 5- or 6- membered rings known as hemiacetals. Forming the hemiacetal (i.e., elimination of the carbonyl) reduces the probability of adding more formaldehyde units.

It is observed that in solution glucose exist more than 99% in the pyranose (i.e., ring) form (36% **alpha** and 64% **beta**, see below). Fructose is an isomer of glucose where the carbonyl is on the number 2 carbon.

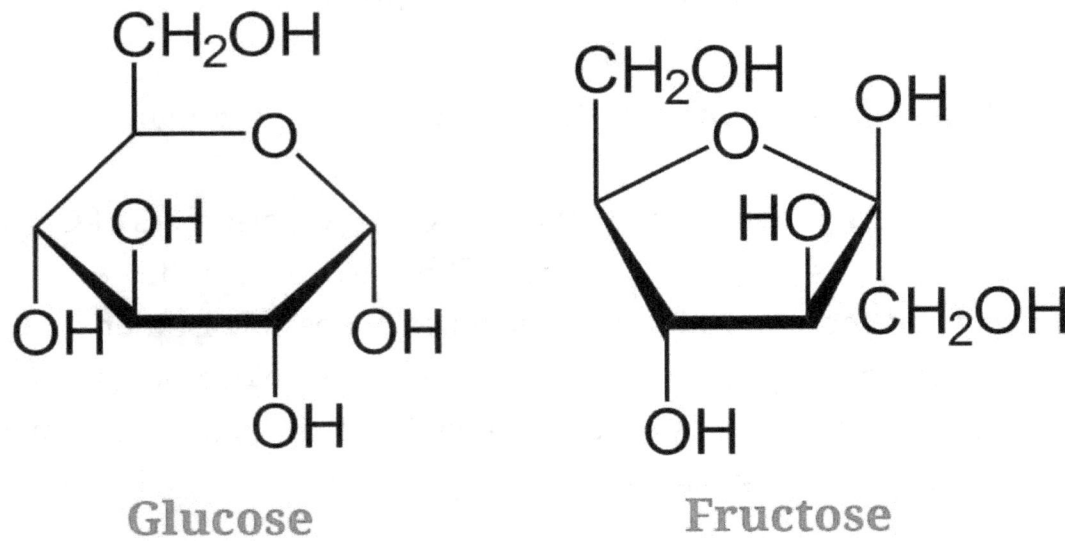

Glucose   Fructose

https://www.nuffzedd.com/wp-content/uploads/2015/12/glucose-fructose.jpg

You will notice that the hemiacetal can from in one of two ways with these molecules (called alpha and beta):

α -isomer   β -isomer

http://www.chem.ucalgary.ca/courses/351/Carey5th/Ch25/ch25-3-1.html

The preferred conformation of glucose is a "chair" form and when the **beta** form is produced in the hemi-acetal both the -CH$_2$OH (carbon 6) and the OH (on carbon 1) are able to be in equatorial positions. In the **alpha** form of the acetal the OH on carbon 1 ends up in an undesirable axial position. In the beta form of pyranose all the -OH substituents can be in the equatorial positions.

Why is the specific Stereochemistry of Glucose Favored?

Why glucose is preferred over all the other possible stereoisomers of C$_6$H$_{12}$O$_6$ is not as great a mystery as might be assumed. It has to do with the stereochemistry and thermal stability of the various possible isomers.
If these isomers are raining out of the sky onto the surface of a hot (100°C to 500°C), moist earth dominated by oxides of metals (sodium, magnesium calcium, aluminum) and metalloids (silicates), hydrolysis and pyrolysis will occur:

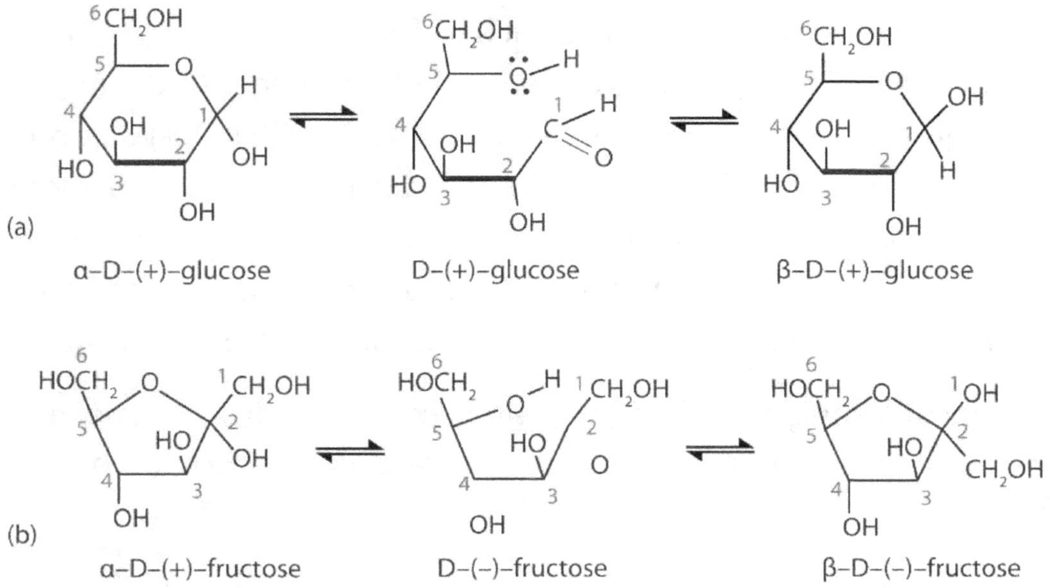

(a)

α–D–(+)–glucose   D–(+)–glucose   β–D–(+)–glucose

(b)

α–D–(+)–fructose   D–(–)–fructose   β–D–(–)–fructose

Source: http://saylordotorg.github.io/text_the-basics-of-general-organic-and-biological-chemistry/s19-04-cyclic-structures-of-monosacch.html

And protonation of the alpha-hydroxide readily leads to levoglucosan[38]:

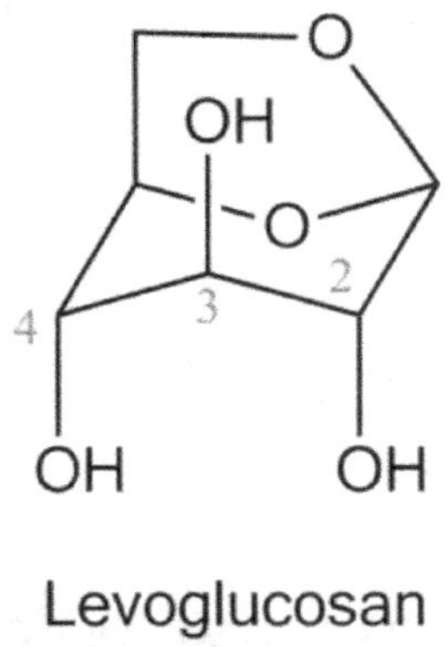

Levoglucosan

**Source:
https://upload.wikimedia.org/wikipedia/commons/thumb/9/91/Chemical_structure_of_levoglucosan.png/220px-Chemical_structure_of_levoglucosan.png**

---

[38] Vikram Seshadri and Phillip R. Westmoreland. Concerted Reactions and Mechanism of Glucose Pyrolysis and Implications for Cellulose Kinetics. *The Journal of Physical Chemistry A* 2012 116 (49), 11997-12013

The point that needs to be made here is that this particular stereo-isomer (of all that might be formed from randomly attached formaldehyde units) is exceptionally thermally stable. It is composed of a 5-membered ring acetal unit with bridgehead hydrogens and a three-carbon (2, 3, 4-CHOH-) chain that is part of a six-membered ring in which the OH units cannot become *trans-periplanar* with the hydrogens on the adjacent carbons. The trans-periplanar arrangement is essential to elimination of H_2O. Thus, the glucose isomer which forms the basis of our biochemistry, was selected by its unique stereo-chemical stability.

In the presence of strong acid levoglucosan is converted levoglucosenone (probably by way of forming a carbocation at carbon #3 followed by elimination):

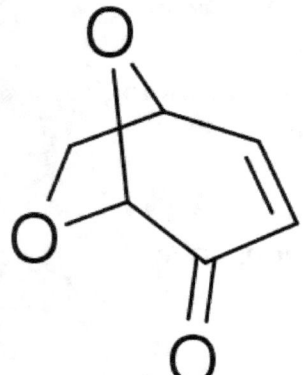

Author: Edgar181; Source: Wikimedia Commons

Polysaccharides

Glucose and ribose were likely two of the most common carbohydrates collecting on the surface of the prebiotic earth. The cracks in the crustal stone likely contained partial hydrolysis products of polyphosphates…polyphosphoric acid (pKa1 = 2.14, pKa2 = 7.20, and pKa3 = 12.37). The acid conditions favored polymerization of glucose[39] producing disaccharides with various α and β

---

[39] Yu Long, Yun Yu, Bing Song, Hongwei Wu. Polymerization of glucose during acid-catalyzed pyrolysis at low temperatures. *Fuel*: 230, 15 October 2018, Pages 83-88.

linkages (including 1,6-, 1,4-, 1,3-, 1,2- and 1,1-glycosidic bonds). The predominant compounds were 1,6-alpha linkages, with the 6-hydroxyl linkage being the most active.

Ribose

The five-membered ring seen in fructose can be formed with polymerization of only five formaldehyde units:

Ribose

These and other sugars (carbohydrates) and glucosamines, thus, accumulated in the water droplets that formed the clouds over the earth's surface. When the surface cooled, they (and products described below) rained down onto the crust. And were concentrated in pools of evaporating water.

But other reactions could take place in the water droplets themselves.

Phosphorylation of Sugars and Related Compounds

An absolutely critical type of reaction for the progression of biochemistry as we know it is the phosphorylation of sugars. This is where the presence of pyrophosphates and phosphosilicates in the mantel is critical. These polyphosphates contain "high energy" P-O-P bonds:

$$\text{P-O-P} + \text{HOR} \rightarrow \text{P-O-R} + \text{P-O-H} + \text{energy (~19 kJ/mole)}$$

Between the time the earth was formed and the time that photosynthesis[40] evolved (a span of billions of years) *this was the only source of energy* to drive organized chemical reaction. Photolysis and electrical discharges generally produced very high energy chemical changes that often destroy more than they build.

It is thus critical that polyphosphates not all be readily and immediately available upon the cooling of the earth and filling of the oceans. If they had been hydrolyzed by the great volume of water, the world would be an ocean filled with phosphate (PO_4^{3-}) and no living thing. Thus, I envision a slow-release of polyphosphate and mineral pyrophosphate from the silicate mantel over time and especially while most of the water was suspended in the atmosphere and the sugars were accumulating on the surface of the earth.

There are two factors that are important about the phosphate esters (P-O-R) that are formed. First, they are quite stable towards hydrolysis. This is unlike for example As-O-R bonds. The relatively small phosphorus atom would require a large promotion of electronic energy to accept an additional nucleophile (-OH) to form a penta-coordinate center, which could expel the -OR group. Thus, the more likely chemical reaction in many cases is the breaking of the O-R with a leaving phosphate ion.

---

[40] Photosynthesis essentially harnessed photolysis to make polyphosphate bonds that had become the essential energy source in biochemistry of life.

$$P\text{-}O\text{-}R \rightarrow PO\text{-} + R+$$

This reaction more often is concerted with a related reaction to avoid excessive buildup of charge on the carbon center or R:

$$P\text{-}O\text{-}R + X \rightarrow PO\text{-} + RX+$$

Second, once complex molecules were formed and subtle changes in the three-dimensional stereochemistry of these molecules became relevant, adding or subtracting phosphate to relatively uncharged -OH groups became important and useful. For example, in a complex molecule, converting a -OH into a phosphate $-OPO_3^{2-}$ can change the preferred conformation of the complex molecule[41]:

$$\textbf{Complex-OH} + {}^{2-}\textbf{PO}_3\textbf{OR} \rightarrow \textbf{Plexcom-OPO}_3{}^{2-} + \textbf{HOR}$$

The change in conformation is driven by interaction with the solvent (water) and other ionic sites within the "complex." Of course, removal of the phosphate can reverse the process. This is the "motor" that runs most of our complex bio-machinery.

2-Deoxyribose

The conversion of ribose to 2-deoxyribose is an interesting problem. There now appears to be an enzyme catalyst (Nelson & Cox, 2000) that preforms a series of otherwise unfavorable and unlikely reactions on the nucleotide. Even with the use of a complicated enzyme, the mechanism proposed by Nelson and Cox involves the undesirable production of a carbocation adjacent to an electron deficient carbon radical:

$$C.\text{-}C+$$

In a world without enzymes, a simpler mechanism comes to mind:

---

[41] Families of *kinase* enzymes have evolved to accomplish these conversions.

Phosphorylation

Formation of a new 5-membered ring

The reduction step seems to be a common sulfide coupling

$$2\ RSH \rightarrow RSSR + H^+ + H^-$$

or reduction with hydroquinone concurrent with the opening of the new (strained) 5-member ring to produce 2-deoxyribose.

I like this mechanism because it is consistent with the stereochemistry and provides a rational for reduction at the 2-position.

Cyanide-catalyzed Benzoin Reaction

It should also be noted that in aqueous solution (e.g., in water droplets in the atmosphere or in puddles of water that collects on the surface of the earth) ammonia (ammonium pKa 9.25) and HCN (pKa 9.21) can react to produce the cyanide anion:

$$NH_3 + HCN \rightleftarrows NH_4^+ + {}^-CN$$

In turn, cyanide anion can react with aldehydes and formaldehyde:

$$RCHO + {}^-CN \rightleftarrows RCH(O-)CN \rightleftarrows RC^-(OH)CN$$

With the generation of a negatively charged carbon that is stabilized by electron withdrawing OH and CN groups. This is the well-known intermediate in the benzoin condensation catalyzed by cyanide:

$$\begin{array}{c} \text{HO}\quad \text{O} \\ 2\,RCHO \rightarrow R\,C - C\,R \end{array}$$

You will notice that this reaction can proceed with addition of additional aldehyde units:

$$\begin{array}{c} \text{HO}\quad \text{O} \qquad \text{HO pika}\quad \text{O} \\ RCHO + R\,C - C\,R \rightarrow RC\text{----}CR\text{ -}CR \end{array}$$

The addition of formaldehyde could continue indefinitely by this mechanism or the aldol mechanism (below) may be involved.

Base-Catalyzed Aldol Condensation

The intermediates containing the $H_2C(OH)$ moiety adjacent to a carbonyl:

$$H_2C(OH)CHO + B^- \leftrightarrows BH + H\text{-}C(OH)CHO$$

are moderately acetic (i.e., the pKa of acetylacetone in water is about 9) and it should not be too hard to extract a proton from carbon adjacent to the carbonyl. If this happens, an aldol mechanism is possible to grow the chain in two-carbon units:

$$H_2C(OH)CHO + H\text{-}C(OH)CHO \rightarrow$$
$$H_2C(OH)CH(OH)\text{-}C^*(OH)CHO$$

But if this reaction were very prevalent, branched chains might be expected, e.g., by condensations involving the carbon marked C*. The stereochemical preference in these molecules may be controlled by coordination to metal ions.[42]

Thus, both on a pH basis and on the basis of the absence of branched chain compounds the aldol mechanism seems to be less likely than the cyanide-catalyzed reactions.

Benzoquinone

The empirical formula of glucose is $C_6H_{12}O_6$ but there are several possible stereo-isomers. Above, it was argued that the particular stereo-chemistry of glucose accounts for its prominence in biochemistry. But the other isomers of glucose do not form such a stable compound as levoglucosan:

---

[42] Kofoed J, Reymond J, Darbre T. Prebiotic carbohydrate synthesis: zinc–proline catalyzes direct aqueous aldol reactions of alpha-hydroxy aldehydes and ketones. Org. Biomol. Chem., 2005; 3:1850-55.

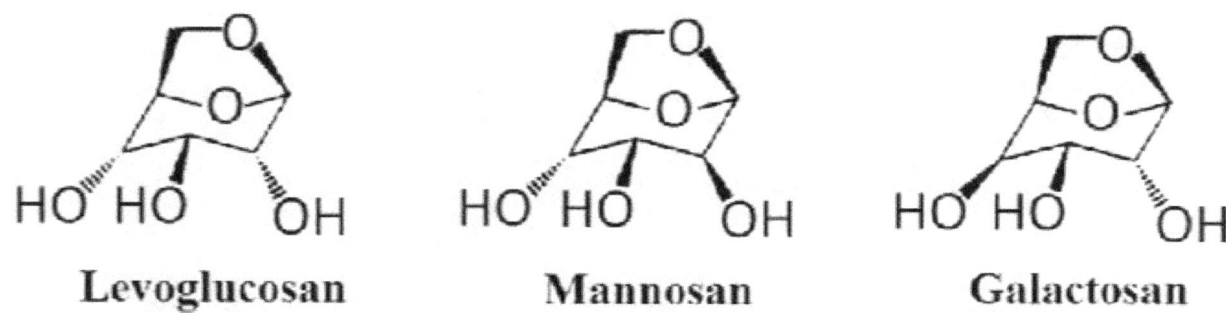

Source: (1) levaglucasone - Bing images

These other isomers may from corresponding (-osan) acetals, but in these cases the elimination of H_2O is much more easily accomplished in *basic or neutral* conditions, which are more likely present on the surface of the earth (resulting from metal oxide reactions with water) One possible mechanism could involve a carbene insertion into a O-C bond to close a 6-membered carbon ring followed by a series of eliminations of H_2O to yield benzoquinone (with some similarity to the Favorskii rearrangement):

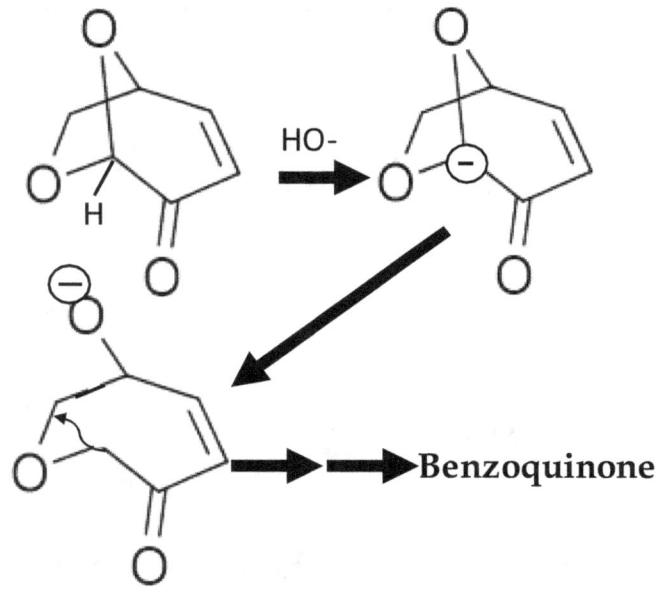

Carbene Insertion

$$C_6H_{12}O_6 \rightarrow C_6H_4O_2 + 4\ H_2O$$

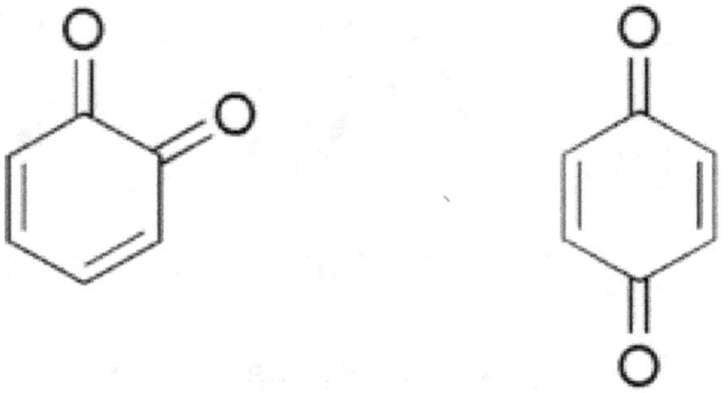

o-Benzoquinone   *p*-Benzoquinone

https://upload.wikimedia.org/wikipedia/commons/thumb/6/66/Benzoquinones.png/320p
x-Benzoquinones.png

Formation of benzoquinone (stable at 500°C and sublimes) from glucose stereo-isomers may have been one of the first (high temperature)[43] processes that occurred when glucose stereo-isomers first reached the surface of the earth.

Benzoquinone is important for participation in oxidation and reduction reactions:

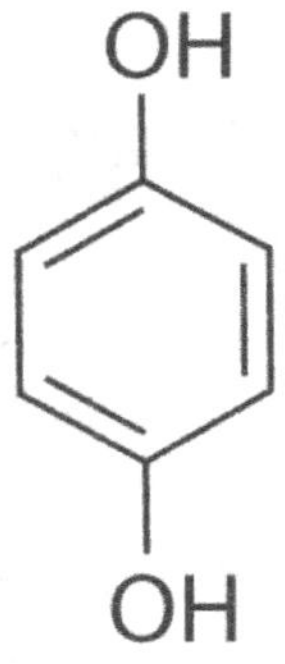

Benzoquinone + 2 H. ⇄ Hydroquinone

In addition, although it is not as obvious how the mechanism would proceed, glucose stereo-isomers and ammonia could undergo condensations, eliminations

---

[43] Stevenson CD, Heinle LJ, Reiter RC. Pyrolysis of p-benzosemiquinone. *J Org Chem*. 2002 Jan 11;67(1):119-24. doi: 10.1021/jo010646p. PMID: 11777448.

and oxidation (by benzoquinone) to produce another important oxidation/reduction electron and hydrogen carrier nicotinamide:

$$C_6H_{12}O_6 + 2\,NH_3 + \text{benzoquinone} \rightarrow C_6H_6N_2O + 5\,H_2O + \text{hydroquinone}$$

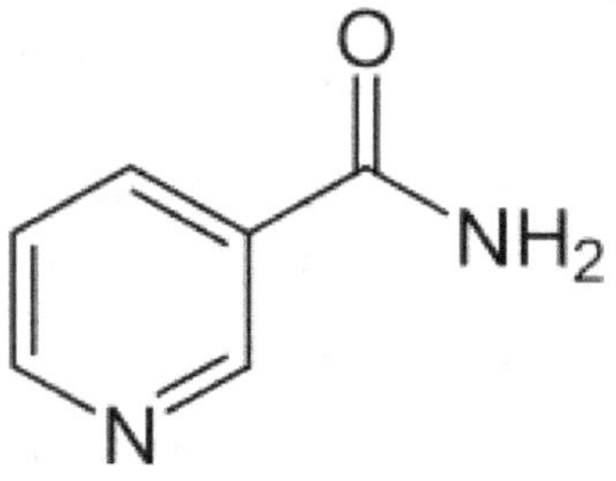

Author: NEUROtiker; Source: Wikimedia Commons

Nicotinamide and form a riboside:

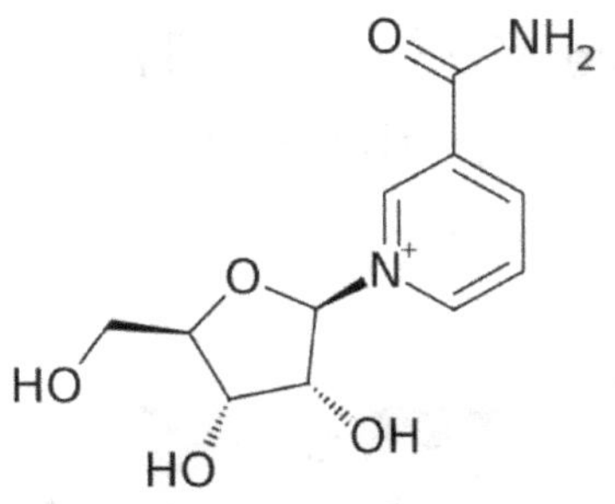

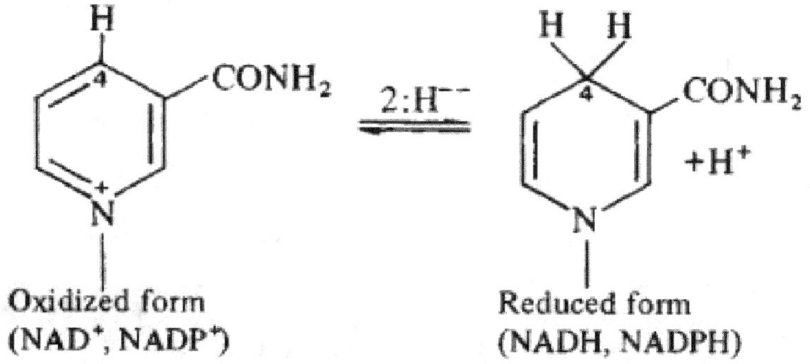

Oxidized form
(NAD⁺, NADP⁺)

Reduced form
(NADH, NADPH)

Source: https://upload.wikimedia.org/wikipedia/commons/thumb/5/50/Nicotinamide-beta-riboside.svg/500px-Nicotinamide-beta-riboside.svg.png

Source: http://img.tfd.com/ggse/26/gsed_0001_0018_0_img4740.png

and is an essential proton/electron carrier in living systems.

Static Electricity as a Source of Free Radicals

The Urey-Miller approach hypothesizes formation of amino acids via lightning acting on methane, ammonia, water and hydrogen (see above). My personal opinion is that the energy involved in a lightning strike is much more likely to be destructive than constructive. I agree that there must have been a long period of near universal rain and storms as the crust temperature decreased from about 500°C to 50°C and water vapor condensed from the atmosphere. But the static electricity that is expressed in lightning strikes begins as charged water droplets in the air. Air is a good insulator and movement of the air causes electrostatic charges to build up in the clouds of condensed droplets. This charge usually is neutralized by coalescence of aerosol droplets of water. But when the cloud extends above the freezing elevation, ice crystals (snowflakes) carrying positive charges tend to stay at the top while water droplets carrying negative charges settle toward the bottom.

I will mention here that the presence of free electrons moving over the surface of water droplets provides a mechanism for absorbing most wavelengths of light. This is similar to the electrons moving across the surface of graphite. The electrons thus absorb every photon that comes their way and the bottom of clouds with electric charge are dark just like graphite. You cannot see the electrons, but there are there soaking up photons.

https://images.freeimages.com/images/large-previews/ed2/storm-clouds-1389802.jpg

Electrical discharges (80% among clouds and 20% from cloud to earth) ultimately equilibrate these charges. The negative charge is probably not carried by water molecules *per se*, but rather by electrophilic molecules dissolved in the water droplets or around the water droplets.

Thus, static electricity (not resulting lighting) provides an alternate and supplemental mechanism for formation of the carbohydrate, benzoquinone and amino acid molecules described above. The mechanisms are very similar but involve free radicals coupling instead of photo-induced 2 + 2-cycloaddition reactions.

In the primordial atmosphere (electrophilic) molecules such as formaldehyde and carbon dioxide are two of the most likely receptors for electrons:

$$[Cloud]e\text{-} + H_2C{=}O \rightarrow H_2C^\bullet{-}O^- \text{ (anion radical)}$$

$$[Cloud]e\text{-} + H_2C{=}NH \rightarrow H_2C^\bullet{-}NH^- \text{ (anion radical)}$$

$$[Cloud]e\text{-} + O{=}C{=}O \rightarrow O{=}C^\bullet{-}O^- \text{ (anion radical)}$$

These intermediates should couple readily with neutral molecules in a polymerization reaction:

$$H_2\overset{\bullet}{C}{-}\overset{-}{O} + H_2O \rightarrow H_2\overset{\bullet}{C}{-}OH + \overset{-}{H}O$$

$$H_2\overset{\bullet}{C}{-}OH + H_2C{=}O \rightarrow \underset{H_2C{-}CH\bullet}{\overset{\overset{\textstyle HO \;\; HO}{|\;\;\;\;|}}{}}$$

$$\underset{H_2C{-}CH\bullet}{\overset{\overset{\textstyle HO \;\; HO}{|\;\;\;\;|}}{}} + H_2C{=}O \rightarrow \underset{H_2C{-}CH{-}CH\bullet}{\overset{\overset{\textstyle HO \;\;\; HO \;\; HO}{|\;\;\;\;\;|\;\;\;\;|}}{}} \quad \text{etc.} \rightarrow \text{carbohydrate backbone}$$

Similarly, we would expect products with the nitrogen analogue:

$$\overset{\bullet}{H_2C} - NH_2 \; + \; H_2C{=}O \; \rightarrow \; H_2C - \overset{\overset{\displaystyle HO \; H_2N}{|\quad|}}{CH\bullet} \; \rightarrow \; [\text{hydrogen abstraction alpha amino alcohol}]$$

$$H_2C - \overset{\overset{\displaystyle HO \; H_2N}{|\quad|}}{CH\bullet} \; + \; H_2C{=}O \; \rightarrow \; H_2C - \overset{\overset{\displaystyle HO \; H_2N \; \; HO}{|\quad|\quad\;\;|}}{CH - CH\bullet} \quad \text{etc.} \; \rightarrow \text{amino-CH}_2\text{O backbone}$$

The intermediate [HOCH$_2$ CH.NH$_2$] can abstract hydrogen or alkyl groups or couple with other groups to form a variety of alpha substituted alpha amino alcohols. These alcohols can be oxidized to alpha amino acids by benzoquinone.

Formamide and urea can also arise by a similar pathway (radical termination step)

$$\overset{\bullet}{H_2C}\text{-}NH_2 + \overset{\bullet}{O}H \rightarrow HO\text{-}CH_2\text{-}NH_2 \rightarrow \text{oxidation} \rightarrow HCONH_2 \text{ (formamide)} + H_2O$$

$$HCONH_2 + NH_3 \rightarrow H_2N\text{-}CH_2\text{-}NH_2 \rightarrow \text{oxidation} \rightarrow H_2N\text{-}CO\text{-}NH_2 \text{ (urea)} + H_2O$$

Lightning: A Destructive Force

Meanwhile, at the surface of the earth, the negative charge in the cloud induces a positive charge in the earth, which concentrates on higher points. The trails of negative ions eventually contact the trails of positive ions and a circuit is completed (cloud to cloud or cloud to ground). Electrical charge passes back and forth along the conducting path caused by the electrons hopping from one molecule to another. This leader current heats the molecules to the point that free electrons and cations can flow. At this point, a massive bolt of charge passes along the conducting path breaking all chemical bonds and actually ionizing some atoms to their core electrons (e.g., stripping C atoms to C^{+6}). As soon as the charge has passed, the electrons and cations begin reforming atoms with the release of photons ranging from microwaves (hence static on the radio) to gamma rays, but mostly in the ultraviolet (which tails into the visible blue region). The thermal temperature of the spark is over 10,000°K (hotter than the surface of the sun). There are typically a series of bolts separated by milliseconds in each discharge event. When the bolts have ended, the ions quickly form neutral atoms and the neutral atoms start colliding to form simple molecules, which collide to form more complex molecules etc. The only thing that really matters to the post-discharge chemistry is what the original of ratios of C:H:O:N were. The main benefit of lightning in formation of oxides of nitrogen and cyanide (NO, NO_2, HCN) from N_2 and CO, which otherwise are not likely to be present in the atmosphere.

2.4 The More-Complicated Building Blocks

Alpha-Amino Carboxylic Acids

Current biology on earth is dominated by proteins, which are assemblages of about 20 naturally occurring alpha-amino acids. The fact that these building blocks use the same basic structure:

$H_2N\text{-}CHR\text{-}CO_2H$

Argues strongly against random recombination of fragments obtained from high-energy processes. Thus, returning to the scheme predicted above the photochemical reactions above the clouds of water droplets would appear to produce exactly the desired combinations. For example, the 2 + 2 cyclo-addition of pi-bonded compounds might have brought us to an intermediate similar to that involved in formation of carbohydrates:

$$H_2C=NH + H_2C=O + h\nu \rightarrow H_2C\overset{\displaystyle O-NH}{\underset{\displaystyle }{\overline{}}}CH_2 \rightarrow HN=CH\text{-}H_2COH \;(\text{minor}) +$$

$$O=CH\text{-}H_2CNH_2 \;(\text{major})$$

This gets us immediately to the universal backbone:

$$\textbf{H}_2\textbf{N-CH}_2\textbf{-CHO}$$

Oxidation of the aldehyde -CHO to a carboxylic acid $-CO_2H$ would easily occur with oxygen. But even without O_2, high energy solar photons are known to produce hydroxyl radicals (HO.) from water, which can directly or indirectly (e.g., via quinones) accomplish this oxidation. Of course, the static electricity discussed above could also induce free radical chain reactions leading to oxidation:

$$\textbf{H}_2\textbf{N-CH}_2\textbf{-CHO} + 2\,.\textbf{OH} \rightarrow \textbf{H}_2\textbf{N-CH}_2\textbf{-CO}_2\textbf{H} + \textbf{H}_2\textbf{O}$$

Glycine

Adding side-chains to the alpha carbon probably involves removing a proton (H+) or hydrogen radical (H.) from the alpha carbon of glycine.

Reaction with formaldehyde yields *Serine*.

The origins of the other side chains are not as obvious but most of them can be envisioned as having precursors subject to nucleophilic or free radical attack by the alpha amino acid moiety (H_2NCHCO_2H). One possibility is that the side chain may originate from a carbohydrate prior to a 2+2 cycloaddition reaction with formaldehyde. In such a reaction path, the carbohydrate intermediates form imines:

$$HOCH_2(CHOH)_nCH=O + NH_3 \rightarrow HOCH_2(CHOH)_nCH=NH + H_2O$$

Which reacts with formaldehyde before or after modification of the side chain. For example (with hydroquinone as a catalyst),

$$HOCH_2(CHOH)_5CH=NH + \text{Hydroquinone} \rightarrow$$

$$\text{Quinone} + 5 H_2O + HO\text{-}C6H4\text{-}CH=NH$$

$$HO\text{-}C6H4\text{-}CH=NH + H_2C=O + \text{Quinone} \rightarrow \text{Tyrosine} + \text{Hydroquinone}$$

Source: Wikimedia Commons

The Purines

Returning to the imine intermediate from formaldehyde and ammonia, purines appear to arise as condensation products of this compound:

$$H_2C=NH + H_2C=NH \leftrightarrows [H_2C^+\text{---}NH\text{---}H_2C\text{---}N^-H] \leftrightarrows H_2C=N\text{---}H_2C\text{---}NH_2$$

$$H_2C=N\text{---}H_2C\text{---}NH_2 + H_2C=NH \leftrightarrows [H_2C=N\text{---}H_2C\text{---}NH_2\text{---}H_2C\text{---}NH] \leftrightarrows$$

$$H_2C=N\text{---}H_2C\text{---}NH\text{---}H_2C\text{---}NH_2 \leftrightarrows \text{ and so forth, followed by cyclization}$$

Notice the alternating C-N units in purines:

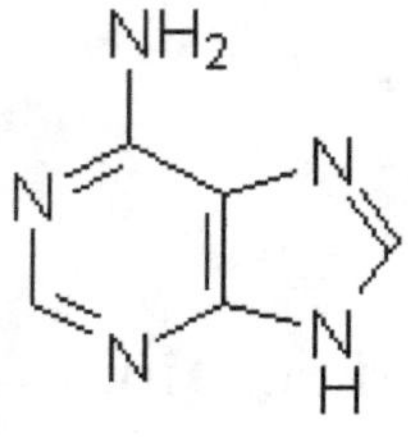

Adenine 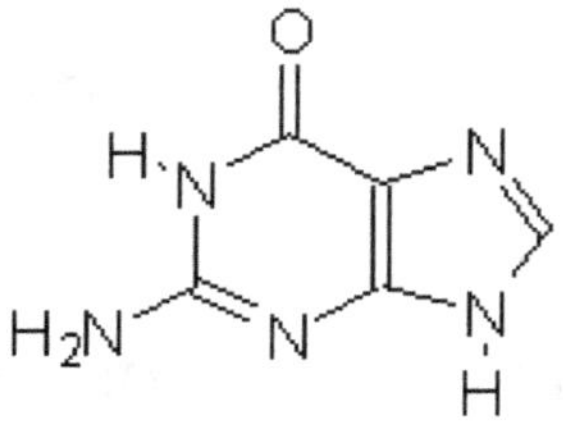   Guanine

Source: http://chemistry.umeche.maine.edu/CHY431/Basics/PurPyrm.html

It is well established that heating formamide produces purines:

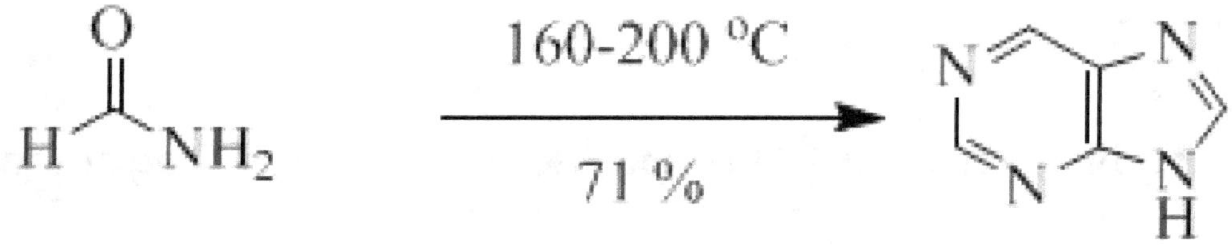

formamide   purine
1

Source: http://en.wikipedia.org/wiki/Purine

The pyrimidines

The pyrimidines likely follow a pattern similar to the purines but include reactions with the carbohydrate intermediate ($O=CHCH_2OH$):

$$H_2C=NH + H_2C=NH \leftrightarrows [H_2C^+\text{—}NH\text{—}H_2C\text{—}N^-H] \leftrightarrows H_2C=N\text{—}H_2C\text{—}NH_2$$

$$H_2C=N\text{—}H_2C\text{—}NH_2 + O=CHCH_2OH \leftrightarrows$$

$$[H_2C=N\text{—}H_2C\text{—}NH\text{—}H(CH_3)C\text{—}OH] \leftrightarrows$$

followed by

(1) oxidation of the C-OH to C=O

$$[H_2C=N \text{---} H_2C\text{---}NH\text{---}(CH_3)C=O]$$

(2) cyclization by aldol condensation

Aldol Condensation

and (3) further oxidation:

Cytosine   Thymine   Uracil

http://chemistry.umeche.maine.edu/CHY431/Basics/PurPyrm.html

Nucleosides (Bases + Carbohydrate)

The labile carbonyl bond in the simple sugars can be turned into imines with the purines and pyrimidines and from there form cyclic compounds called nucleosides:

R-CHO + H-Nbase ⇆ R-CH(OH)Nbase ⇆ Nucleotide

Adenosine	Guanosine
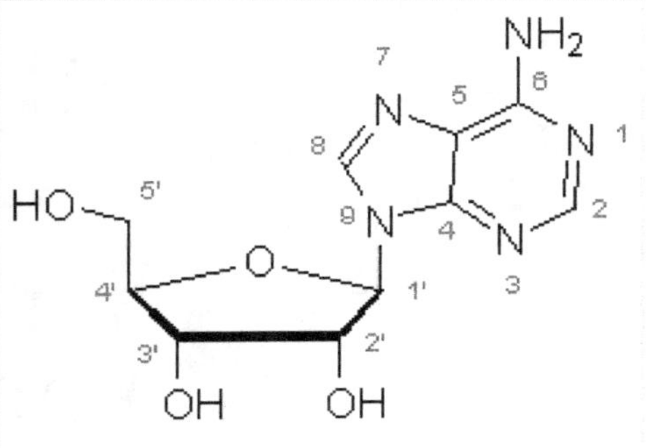	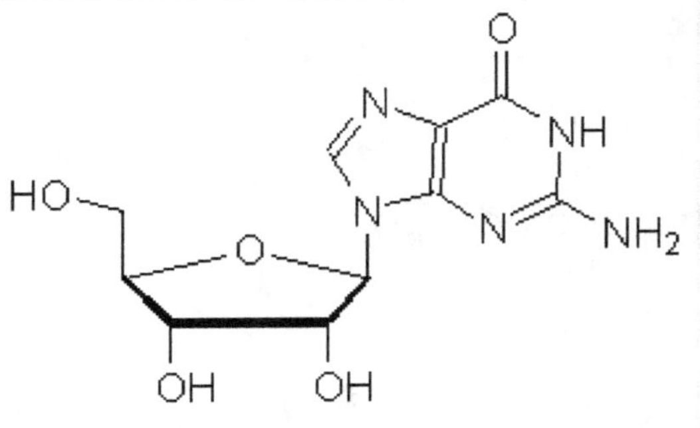

Source: http://chemistry.umeche.maine.edu/CHY431/Basics/Nucleo.html

Cytidine	Uridine
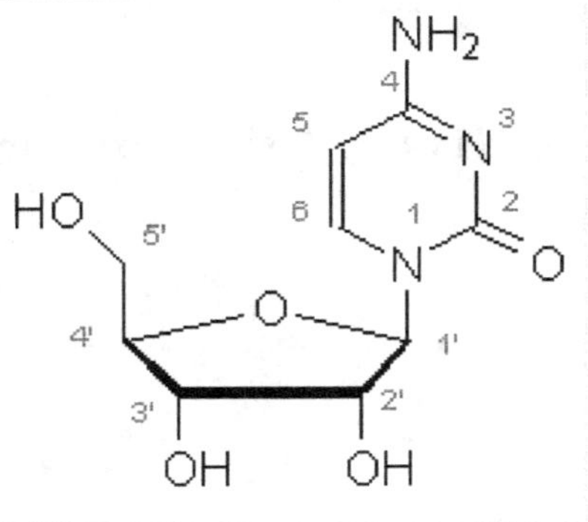	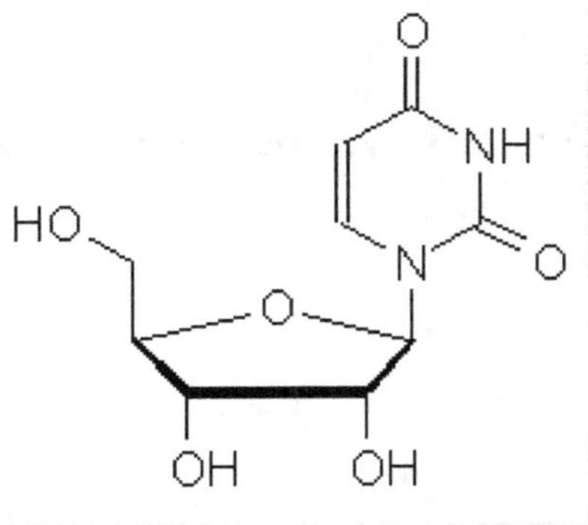

Nucleotides

The 5′ hydroxyl groups on the nucleotides are readily accessible to formation of phosphate esters. Encounter of an adenosine nucleotide with a trimetaphosphate (see above) will likely immediately produce a compound we know as ATP:

Source:
https://upload.wikimedia.org/wikipedia/commons/thumb/f/f7/ATPanionChemDraw.png/12 00px-ATPanionChemDraw.png

This molecule is the most ubiquitous source of energy (chemists would say "free energy") for making otherwise unfeasible reactions occur.

$ATP + H_2O \rightarrow ADP + P_i$ $\Delta G \approx -31 \text{ kJ/mole}$

$ATP + H_2O \rightarrow AMP + 2\,P_i$ $\Delta G \approx -61 \text{ kJ/mole}$

$2\,ADP + H_2O \rightarrow 2\,AMP + 2\,P_i$ $\Delta G \approx -61 \text{ kJ/mole}$

Cyclic Nucleotide Monophosphates and RNA

Cyclic adenosine (5′-3′) monophosphate has survives in our metabolism

Author: Smokefoot; Source: Wikimedia Commons

Its existence points to an era in which reaction of nucleoside triphosphates with themselves released diphosphate and cyclic monophosphate nucleosides. It does not take much imagination to envision a chain reaction of sorts in which randomly sorted clusters of cyclic monophosphate nucleosides zipped together to form polymers we would recognize as ribose nucleic acids (RNA).

In 1986, Orgel[44] published an interesting manuscript abstracted as follows:

> *It is proposed that mononucleotides incorporated into the surfaces of microcrystals of inorganic phosphates such as hydroxyapatite can act as templates to assemble complementary mononucleotides from solution, and that the phosphate groups of the assembled nucleotides can facilitate nucleation of a second hydroxyapatite*

---

[44] Orgel, L.E. Did template-directed nucleation precede molecular replication?. *Origins Life Evol Biosphere* 17, 27–34 (1986). https://doi.org/10.1007/BF01809810.

crystal. This would provide a mechanism of replication that is subject to natural selection. The possible role of a replicating system of this kind in the origins of life on the earth is discussed.

Abiotic Polypeptides (Polyaminoacids)

It is possible that the first polyamides from amino acids were actually formed by oxidation of imines[45] (see the synthesis of pyrimidines above):

$$H_2N\text{-}CHR\text{-}CH=[=N\text{-}\textbf{CHR}\text{-}\textbf{CH}=]n=N\text{-}CHR\text{-}CHO \rightarrow \text{oxidation} \rightarrow$$

$$H_2N\text{-}CHR\text{-}CO\text{-}[\textbf{NH}\text{-}\textbf{CHR}\text{-}\textbf{CO-}]n\text{-}NH\text{-}CHR\text{-}CO_2H$$

Alternatively, mixed carboxylate-phosphate anhydrides would do the job.

$$H_2NCHRCO\text{-}O\text{-}Pi + H_2NCHRCO_2H \rightarrow H_2NCHR\text{-}\textbf{CO}\text{-}\textbf{HN}\text{-}CHRCO_2H$$

Of course, the possibility of raw thermal dehydration of amino acids to yield abiotic polypeptides should not be ruled out. [46]

---

[45] Jan Larsen, Karl A. Jørgensen, Dorthe Christensen. Duality of the permanganate ion in the oxidation of imines. Oxidation of imines to amides. *J. Chem. Soc., Perkin Trans. 1*, 1991, 1187-1190.

[46] Fox SW, Harada K. Thermal copolymerization of amino acids to a product resembling protein. *Science.* 1958 Nov 14;128(3333):1214. doi: 10.1126/science.128.3333.1214. PMID: 13592311.

Harada K, Fox SW. Thermal synthesis of natural amino-acids from a postulated terrestrial atmosphere. *Nature.* 1964 Jan 25;201:335-6. doi: 10.1038/201335a0. PMID: 14109988.

Abiotic Peptides (Proteinoid Microspheres)

Sidney Walter Fox (1912–1998) appears to have been the first person to suggest that thermal peptides might produce protocells.[47] These micro-spheres have several remarkable properties:

(1) They form a shell with an interior volume of water that is partially isolated from the bulk water

(2) They accumulate more peptides into the shell and the sphere grows

(3) When the sphere reaches a certain size, thermodynamics make it unstable relative to smaller spheres and the large spheres fission into smaller spheres.

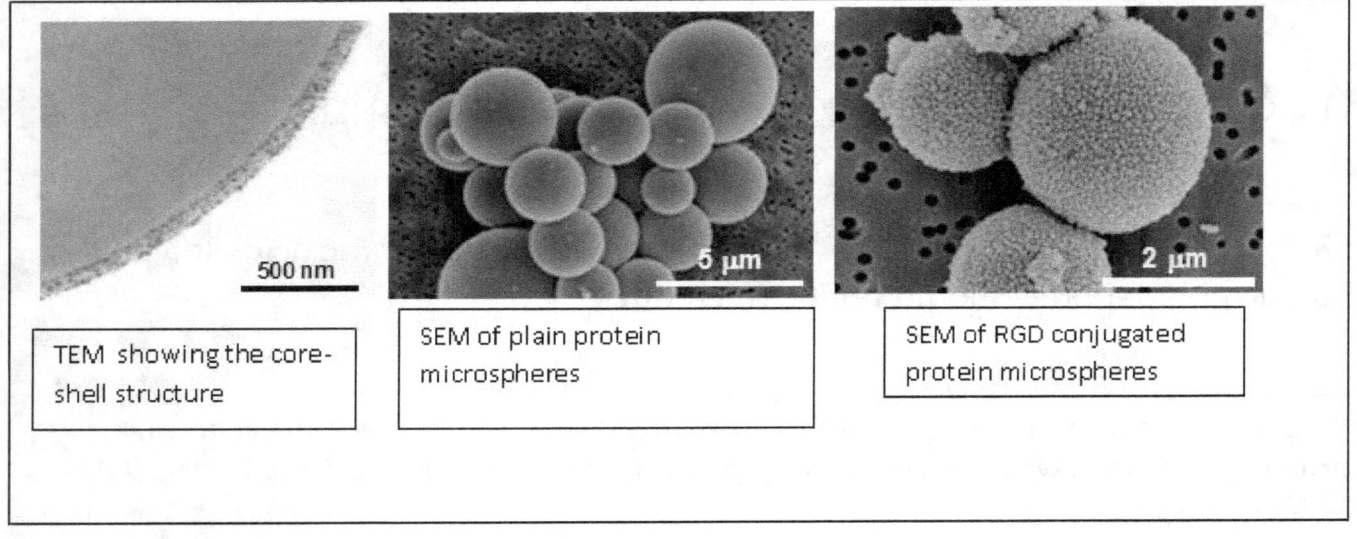

500 nm

5 µm

2 µm

TEM showing the core-shell structure

SEM of plain protein microspheres

SEM of RGD conjugated protein microspheres

Source: https://www.biophotonics.illinois.edu/images/mi4.jpg

---

[47] Hsu LL, Brooke S, Fox SW. Conjugation of proteinoid microspheres: a model of primordial communication. *Curr Mod Biol*. 1971 May;4(1):12-25. doi: 10.1016/0303-2647(71)90003-7. PMID: 5577091.

In a series of papers, Fox and coworkers developed and expanded the ideas that were actually an extension of the Urey-Miller assumption of a protein-based beginning of life on earth:

> [48]*Proteinoid microspheres, produced under geologically relevant conditions, have been found to form junctions. Some of these junctions are hollow; internal particles much smaller than the original microspheres transfer between microspheres. Since these particles contain macromolecular information, they represent a model of primordial communication and of inheritance. The phenomena observed have been viewed as an imposition of constraints on Brownian motion.*

> [49]*Proteinoid microspheres with stable internal compartments and internal structure are made from acidic proteinoid and basic proteinoid with calcium. The populations of microspheres are characterized by a wide diversity of structure. A model of primitive intracellular communication is suggested by the observed movement of internal particles between compartments of a multicompartmentalized unit. Differential response to pH change and to temperature change has been demonstrated within one population and suggests one mode of adaptive selection among primordial cell populations.*

An Important Conclusion: Biochemistry is Universal

At first glance, biologically important molecules such as amino acids, purines and pyrimidines, and sugars appear to be complex chemical structures and it is easy to see why non-chemists could be skeptical about the production of these compounds from "air" without some Devine guidance. On the other end of the

---

[48] Laura Ling Hsu, Steven Brooke, Sidney W. Fox. 1971. Conjugation of proteinoid microspheres: A model of primordial communication, *Biosystems*. 4 (1):12-25 1971,

[49] Brooke S, Fox SW. Compartmentalization in proteinoid microspheres. *Biosystems*. 1977 Jun;9(1):1-22. doi: 10.1016/0303-2647(77)90028-4. PMID: 20178.

spectrum, science fiction writers may imagine that on other worlds, quite different molecules would form the basis of life. Actually, given the conditions of the early earth (which are similar to the conditions that will be found in many other solar systems) the molecules that life on earth is based on are easy to obtain while other possible molecules are quite unlikely.

The conclusion is that once a planet like earth is formed, the building blocks of life as we know it are almost certain to form and we can expect that these same molecules will dominate the biochemistry of alien planets. By extrapolation, we can expect to find that life on other planets will likely be based on the universal building blocks found here although organisms will evolve differently.

What is NOT Here

One class of compounds that you may notice is missing from the analysis above is lipids. The formation of hydrocarbons with extended chains of $-CH_2-$ is not likely until photosynthesis is available. At that point, polyolefins ($-HC=CH-$) will be found to collapse into well-known 5- and 6-membered rings (e.g., steroids, terpenes) and form fatty acid and fatty acid esters. The enormous energy provided by photosynthesis and some rather specific chemical processes are needed to accomplish these transformations.

Forward to Part 2

In the paragraphs above, I have tried to outline what I think are the most important elements of chemistry leading to a variety of molecules from which life can evolve. I strongly differ from the widely accepted Urey-Miller for a variety of reasons primarily the composition of the primordial atmosphere and the secondarily the relevance of electrical discharges in facilitating important chemical reactions. Most work (e.g., Sidney W. Fox) has followed from the assumptions of the Urey-Miller experiments and focused on proteins. I think that is rather short sighted.

In Part 2, I propose to summarize and discuss *RNA World* and the advent of living cells. From there, evolution (by natural selection) can proceed. The establishment of the DNA-based system of genetics is a major milestone in the progress of life on earth.

The next big event is evolution of photosynthesis. Although this is long after the establishment of life on earth, I think it is worth briefly discussing the adaptation of life to the constant bombardment of high energy photons from the sun.

I have already published hypotheses regarding the current mechanisms of evolution and development.